THE ENCHANTED VOYAGE AND OTHER STUDIES

Frontispiece. William Bateson, F.R.S., discoverer with R. C. Punnett of the phenomenon of genetic coupling and repulsion, now called linkage, and the basis of the whole of modern genetics. (Crayon drawing by D. G. Lillie, 1909.)

G. E. Hutchinson

THE ENCHANTED VOYAGE AND OTHER STUDIES

GREENWOOD PRESS, PUBLISHERS
WESTPORT, CONNECTICUT

Library of Congress Cataloging in Publication Data

Hutchinson, George Evelyn, 1903-
The enchanted voyage and other studies.

Reprint of the 1962 ed. published by the Yale University Press, New Haven.
Includes index.
1. Biology--Addresses, essays, lectures.
I. Title.
QH81.H898 1978 574'.08 77-26010
ISBN 0-313-20098-X

ERRATA

On page 134, line 5 should read *church having been rebuilt in the baroque style in the eighteenth century, under a side*

On page 134, line 24, the word *monk* should be *abbe.*

The name *Florence* should be *Winifred* on the following pages:

page 6, footnote 7
page 13, line 10
page 157 (index)

Reprinted with the permission of Yale University Press

Reprinted in 1978 by Greenwood Press, Inc.
51 Riverside Avenue, Westport, CT 06880

Printed in the United States of America

LAVRAE MANGIONE

PANORMITENSI

DOMINAE CHRISTIANAE DOCTISSIMAE

AVCTOR

DD

PREFACE

Nine of these essays, including the six here presented as an historic sequence under the title "Dissolving Views," first appeared in the *American Scientist.* Two others were published in the *American Naturalist*, and one apiece in the *Journal of Marine Research*, *Curator*, and the *Yale Review*. I am grateful to all these journals for permission to reprint. The three remaining studies, "Pascà," "Meditation on Death in the Churches of Florence," and "Cambridge Remembered," have not been previously published. The first two of these are in a sense parerga constructed of memories, partly gathered during a Guggenheim Fellowship awarded for quite different studies. It would be invidious therefore not to express on this, as on other occasions, my appreciation of the John Simon Guggenheim Foundation, surely the most thought-provoking institution in America. Almost all the previously published pieces have been extensively revised.

My thanks to George A. Baitsell, formerly editor of the *American Scientist*, originally expressed in the predecessor of the present volume, are reiterated here. My particular debt to Signorina Mangione is recognized on another page. A number of individual acknowledgments for specific corrections or pieces of information will be found in the notes appended to the various essays. My wife remains my most perspicacious critic, and this volume owes more to her than I can express.

In reading one of the more recently published pieces, along with other, largely political and literary articles, in a periodical, I have realized that in their incessant preoccupation with strange and rare animals and plants, the disappearing traces of obsolete religious customs, odd neglected works of art, and the purely aesthetic aspects of scientific theory, these studies must appear to those concerned with international politics and the literary expression of anxiety as incurably frivolous. This is in a sense their initial intention. The really important human problems probably cannot be stated in any clear way, much less can any formal solution be offered. That the universe and its inhabitants can be extremely decorative is easily forgotten, particularly today, but the subjective realization of this decorativeness is presumably part of reality and may provide one way of dealing with our problems obliquely, in the end, which we cannot reach, leading to the exclamation of the Easter antiphon: "Such knowledge is too wonderful and excellent for me. Alleluia."

Michaelmas 1961
New Haven, Connecticut

G. E. HUTCHINSON

CONTENTS

ILLUSTRATIONS

THE ENCHANTED VOYAGE[1]

As a beginning, an examination of the first scene of *The Tempest* will be appropriate, since it introduces us to a world where terrestrial action and maritime action are in conflict. The situation reminds us of one of those extraordinary paintings of Magnasco in which a seascape seems to be over-powering a landscape.

The Tempest opens with the members of the royal household of Naples, who have authority on land, abusing the boatswain, who has authority on the ship. However, the old councilor Gonzalo, who is better and wiser than the other landsmen, expresses his confidence in the boatswain with these very remarkable words: "I have great comfort from this fellow: me thinks he hath no drowning mark upon him, his complexion is perfect gallows. Stand forth good Fate to his hanging: make the rope of his destiny our cable . . . " It is clear from this that the predicament in which the passengers found themselves would not end fatally.

T. S. Eliot, in *The Wasteland*, his first major work, used, as is well known, a number of images and phrases from *The Tempest*, Shakespeare's last work. In a certain sense *The Wasteland* is the converse or reciprocal of *The Tempest*. Of the images that Mr. Eliot rejected, at least initially, the hanged man is pre-eminent. Madame

1. Given as an evening lecture at an Oceanographic Convocation held at Woods Hole, Massachusetts, June 1954, on the occasion of the dedication of the Laboratory of Oceanography, established by the Office of Naval Research.

Sesostris, the fortuneteller with a somewhat unorthodox pack of tarot cards, remarks, "I do nót see the hanged man. Fear death by water." In the fully developed lore of the tarot cards, the hanged man is given the meaning of the death of self as a prerequisite for illumination. Little enough is known of the origin of the tarot; it is not unreasonable to ally the hanged man to the Crucifixion but this can hardly be a primitive meaning. Whatever the original interpretation may have been, Eliot consciously rejected the symbol, substituting for it the drowned man, Phloebas the Phoenician sailor, though, as his notes imply, the hanged man appears late in the poem in a shadowy form. For Eliot, the magical storm, which is innocuous in *The Tempest*, causes a disaster, and although all is not lost, the king is left fishing in a squalid world with an arid plain at his back, considering how he may salvage the little that is left of his possessions.

It is my purpose to show that an antithesis like that between the hanged man and the drowned man of *The Tempest* and *The Wasteland* has behind it much that is magical in the sea.

We may regard any oceanographic expedition as one sort of voyage, having some of the general properties of other voyages. There are a number of men in a ship. Traditionally, they are separated from women; they could not form a biologically stable community. Like any of the oceangoing sea birds, man finds that he must return to the land to breed. Thus there is always the danger that the men on a voyage may be stranded on Circe's domain, or encounter the Sirens, or run aground on any of the numerous islands of women that are so persistent a feature of mythology.[2] Just as the modern expedition is in search of knowledge, so its mythological prototype is in search of magic. The voyage is undertaken to make the sea yield enchantment, which, though dangerous, may also be useful and beautiful.

All objects of the search will be things that still might be within the domain of any expedition studying biological oceanography.

2. Anthony Thorne's *Young Man on a Dolphin* (London, William Heinemann, 1952) provides a recent and very delightful example.

There will be time to consider only a few coelenterates and mollusks, precious coral, cowries, the sea snails that yield Tyrian purple, and pearls. One might add many other species, and one should not be surprised at the presence in the collection of quite unlikely organisms. For instance, there was probably an *Octopus* cult in the Mediterranean in prehistoric times, and this seems to have spread as far northwest as the island of Alderney.[3] It will not be possible, however, for one investigator to examine the whole collection, hence much of interest must be put to one side. Before beginning our study, it is desirable to remember that there is a persistent tendency to people the sea with beings paralleling those on land. When Alexander the Great is depicted in a medieval miniature of what Dr. Beebe has himself recognized as a prototype of the bathysphere, he is shown in a submarine landscape containing not only fishes but a thicket of submarine trees and a pair of submarine human beings, a merman and a mermaid. Today we have sea horses, sea cows, sea dogs (*canes marini* or dogfish, of which more hereafter), sea robins, sea urchins, sea anemones, as well as a few parallels of human artifacts such as sea pens and sea squirts. It will be clear, in fact, that the drowned man is in part a sea hanged man. For the moment, however, it may merely be noted that belief in such correspondences, which, incidentally, is mentioned in a more recondite form in the prefatory note to Shelley's *Sensitive Plant*, was very widespread.

The first specimens collected on our Enchanted Voyage belong to *Corallium rubrum*, the precious coral. In many early Italian paintings of the Madonna and Child, the latter wears at his throat a small branch of coral. In a few paintings this is accompanied by some other ornament which may even be a cross. The coral is sometimes a natural branch, somewhat polished; in other examples it may be worked into the shape of a horn. Such coral horns are still worn

3. T. D. Kendrick (*The Archaeology of the Channel Islands*, Vol. 1, *The Bailiwick of Guernsey*, London, Methuen, 1928) figures a problematic stone object from a megalithic site on Alderney with the tentative suggestion that it is a cephalopod. There is no doubt that it represents *Octopus* or *Eledone*.

widely in southern Italy, where it is believed that they protect the wearer against the evil eye. Since children are considered to be particularly sensitive to such a baleful influence, it is reasonable that they should wear this ornament. In pagan times every Roman baby had hung about its neck, for the same purpose, a frankly phallic charm called the *res turpicula*, which was sometimes, though not invariably, made of coral. Comparable coral phallic pieces are perhaps still worn, but not conspicuously, by adults.

There is also known a very remarkable charm which was used by pregnant women and which was hung at the necks of very small children.[4] This charm is termed the *cimaruta* (Plate 1, *below*) or *fogliaruta*. It seems to have been used mainly around Naples, and by 1925 was becoming obsolete, only very degenerate examples being made. There are no specimens in the Museo Pitrè, the great collection of Sicilian folk objects in Palermo, and the charm appears unknown in Sicily. The silversmith near the Badia in Florence still has examples of good workmanship that he has made, but I think these are sold purely as ornaments. Modern copies, with inset paste or semiprecious stones, have also been available in other parts of northern Italy[5] as costume jewelry.

The charm is composite; in fact, some of the elements could be obtained separately. The standard components are a crescent moon, usually two birds, a key, and a flower which is supposed to be the rue (*Ruta graveolens*) flower, though it lacks the stamens that alternate with the five petals.[6] An essential part of the charm is evidently

4. The great collection of these objects, that of Elworthy, is in the museum at Taunton, Somerset, England. See also F. T. Elworthy's *The Evil Eye* (London, J. Murray, 1895), from which work some of my information is derived. However, Elworthy did not realize the connection between coral and the cimaruta, though he figured the bronze object at Bologna.

5. Signora Livia Pirocchi Tonolli's perennial kindness for this and many other things is gratefully acknowledged.

6. Many rue flowers are in fact quadriradiate rather than quinquerradiate. Rue was to Pliny (*Nat. Hist.* XX.51) "in praecipuis . . . medicaminibus," and he gives so many uses that it is impossible to trace any consistent scientific or magical connections.

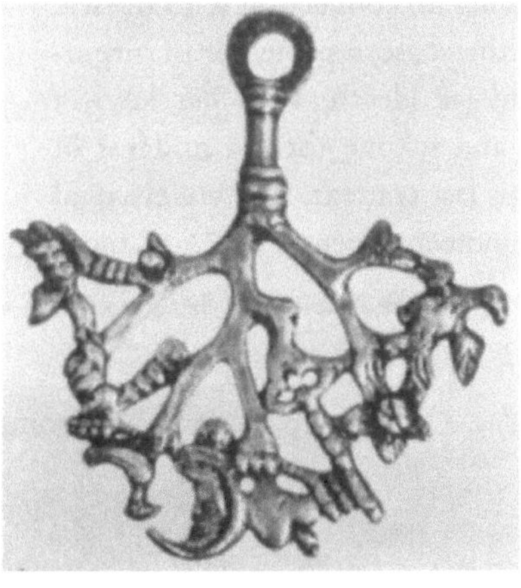

PLATE I. *Above:* Christ child (detail) from Piero della Francesca, "Madonna and Angels," in the Ducal Palace in Urbino (Alinari photograph.)

Below: A cimaruta, probably early nineteenth century, Naples.

one or more clenched hands, with the thumbs projecting between the index and middle fingers. This obviously sexual gesture, protective against *malocchio* and known as *mano a fico*, is also found on the Roman *res turpicula*. A modern cimaruta adds a snake, presumably after some old variant. In all good old examples, from the Kingdom of the Two Sicilies, the ring at the top and sometimes other parts of the charm bear the hallmark "N A P"; it was considered important that the cimaruta be made of silver. There is little doubt that the moon, key, and perhaps the rue flower are connected with Artemis or Hecate; the requirement that the charm be of silver fits a moon goddess also. However, when we extract these elements, and the birds, and *mano a fico*, we are left with a branching structure which only vaguely resembles a rue leaf, still less a sprig of the plant, but it is almost identical with an early Iron Age object in the Bologna Museum which is merely a bronze model of a branch of coral. Any doubt as to the identity of the basic structure of the cimaruta and a piece of coral is dispelled by the coral ornament worn by the Christ Child in the Piero della Francesca "Madonna and Angels" in the Ducal Palace at Urbino (Plate 1, *above*). Pliny tells us that coral was useful medicinally in infancy and that it also prevented nosebleeding. It may be significant that, within the past half century, pregnant women who were unable to afford a silver charm might stuff actual rue sprigs up their nostrils.[7] It would seem reasonably clear that the coral and the rue are to some extent interchangeable in prophylactic magic; from this point of view coral is a kind of sea rue, or perhaps rue a kind of terrestrial coral.

As ordinarily understood, Artemis has no connection with the sea. She is a goddess of the moon and of the chase, a protector of virginity, a guardian of the nether world (as Hecate with her key), a patroness of women in childbirth, and in one form a goddess of marshes. The frog, her symbol in this last context, was widely used as a charm, at least until the late nineteenth century, if not more

7. This information was given the writer in 1926 in Naples by Miss Florence Allen.

recently. A frog, however, is essentially a freshwater animal. The association of Artemis with coral is therefore not easily explained and can hardly be primary.

Another very common Neapolitan charm is in the form of a mermaid, often with two tails, or of a woman riding on a sea horse. Such a charm was called *sirene*, so we now have caught another mythological being out of the sea. Though the purpose was the same as that of the cimaruta, the sirene was hung in a house or formed part of the elaborate ornaments of the harness of horses.[8] One remarkable feature of the sirene is that it was usually ornamented with bells. As Elworthy points out, similar bells formed part of an object called *coral and bells*, which was customarily given to teething infants in England, at least down to the middle of the last century. It consisted of a branch of coral with a rattle bearing the bells and a whistle at one end. The coral and bells seems to link the belled sirene with the coral horn and the silver cimaruta. There may be a connection, though a remote one, between such objects and the curious ornament, consisting of a branch of coral, at the ends of the ramifications of which there are either pearls or silver bells, worn on the front of the head of one of the two angels in the central panel of the Portinari altarpiece of Hugo van der Goes in the Uffizi.[9] That all these charms are complex and inconsistent probably indicates a long and elaborate history. The association with Artemis or

8. Traditionally this was most elaborate; the writer has seen the central part swathed in black during Holy Week. Elworthy describes some of the objects involved, but does not explain a peculiar and moderately common figure, that of a boy wearing a conical cap holding his shirt up to his waist in front. This motif was still apparent in 1958. Made up in china with a rubber cap, the same figure was sold as a vulgar toy at *feste*, but from its appearance on horse trappings it is practically certain than an indefinite mythological vista, obviously involving Attis, as Alan Little pointed out in conversation long ago, lies behind what might otherwise be regarded merely as a childish impropriety. All these saddle ornaments seem now (1961) to have become obsolete.
9. The writer is indebted to George Kaftal for pointing this out.

the Sirens may well be a late aspect of that history. All that we can rightly conclude is that they involve some mythological lady out of the sea.

Of all such mythological ladies, none is more justly celebrated than Aphrodite, and when we turn to her we find the most odd situation. Hesiod, in his *Theogony*, says she arose from the sea foam. The ancient Greek philologists indeed derived her name from ἀφρός, but they were incorrect according to modern researches. She is usually thought to be of Eastern origin; the Greeks themselves supposed she came from Cyprus. Hesiod's story is a very strange one. He related that Kronos, the son of Uranos, castrated his father and that the foam from which Aphrodite arose collected round the severed genitalia of the old sky god which had fallen into the sea. There was another story of the origin of Aphrodite, given by Homer, that puts her much later in celestial history as the daughter of Zeus and Dione. The later Greeks, in fact, distinguished between the motherless Aphrodite Urania and the more conventionally generated Aphrodite Pandemos, but there is no ancient authority for this. Cornford,[10] among others, has clearly shown that Hesiod's *Theogony* is derived from the ancient mythologies of the mainland of western Asia; such mythologies are apparently explanations of still older rituals. We may guess vaguely that behind the succession of Uranos, Kronos, and Zeus lies a series of divine kings, mutilated or killed by their successors, and that the generative power of water and the abundance of the sea was somehow supposed to depend on such bloody practices.

Of all the objects that can be associated with Aphrodite, none is more significant than the cowrie shell.[11] We use her name *Cypraea* for the type genus of the cowrie family. Egyptian princesses, far before the Greeks existed as we know them, wore magnificent

10. F. M. Cornford, *Principium Sapientiae* (Cambridge Univ. Press, 1952).

11. See particularly J. Wilfred Jackson, *Shells as Evidence of the Migrations of Early Culture*, with an Introduction by G. Elliot Smith (Publ. Univ. Manchester, Ethnog. Ser. 2, 1917).

girdles of gold cowrie shells which doubtless promoted fertility. In Ladakh or Indian Tibet the women still wore, at least until a few years ago, an elaborate ornament composed of a brass fret from which hung strings of the money cowrie *Monetaria*; this fret was given a name that specifically implied a visceral significance; the word for a cowrie appears also to mean womb. It was worn on the right hip, Gautama Buddha having been born from the right hip of his mother. When a woman was wealthy enough, she would add further strings of cowries to her belt and might attach some of the large species of *Cypraea* in addition.[12] The cowrie is also still used as a charm in Arab countries. In the Museo Pitrè there are two specimens of *Cypraea lurida* and one of a larger, and apparently not Mediterranean, species that had been used in the last century in Sicily, as had a walnut and a pierced horse chestnut, which were possibly cowrie substitutes.

If we now return to land and look for the terrestrial parallel of the cowrie, we find it in the mandrake, the true Old World mandragora, as the evidence collected by Rendell Harris [13] clearly shows. Mandrake roots were used in a way like cowries, hung around the waist to promote fertility. At this point, very unexpectedly, the hanged man enters, for the mandrake was believed to grow at the foot of a gibbet on which a man, preferably innocent, had been hung.[14]

To pick a mandrake was very dangerous, in fact fatal, so the medieval herbalist recommended that the plant be loosened and then tied to the tail of an unwanted dog. The dog was to be enticed from the spot with a piece of meat. As it ran to get the meat, the mandrake

12. G. E. Hutchinson, *The Clear Mirror* (Cambridge Univ. Press, 1937). See also "Lakes in the Desert," *American Scientist*, *37*, (1949), 399, for a photograph of a western Tibetan lady wearing the ornament in question.

13. See *The Ascent of Olympus* (Manchester Univ. Press, 1927).

14. It was apparently said that the plant was derived from the semen of the corpse, which completes the Uranos parallel quite diagrammatically. It is also interesting and significant that psychoanalysts believe that fear of the evil eye is based on a repressed fear of castration.

was pulled out, giving a horrible shriek as it left the earth; the unhappy dog expired and the herbalist went off with his prize. The linking of a dog to the terrestrial expression of Aphrodite is very interesting, because it is reflected in the extraordinary story of the origin of Tyrian purple. This dye, a bromine derivative of indigo, is produced in the pallial glands of a number of species of *Murex* and *Thais* and perhaps of other genera of the *Muricidae*. It was used as a dye in various parts of the Old and the New World. One may wonder how the process was ever discovered, particularly since the dye is present in the gland as a leuco-form which requires exposure to air and light for some minutes before the purple colour develops. The ancients, however, knew that one day Hercules was walking with a young Tyrian lady along the beach. His dog bit at something and then began to foam purple at the mouth. The lady was so intrigued with the color that she challenged Hercules to obtain her a dress of the same shade. Hercules, being a practical man, went to the source of the matter and established the Tyrian dye industry. There is a very persistent myth about sharks, *canes marini*, guarding pearls; in some cases in which the pearl fisher is killed by the shark, it seems possible that we have a story of which the mandrake myth is a confused and inverted version.[15] The pearls are certainly connected with Aphrodite and they are, as both Shakespeare and Mr. Eliot assure us, the drowned man's eyes.

We may conclude, therefore, from this brief examination of part of the collection amassed on the Enchanted Voyage, that, thousands of years ago, when ships first put out into the Persian Gulf and the seas about Egypt, many of the objects that could be hauled up were endowed with magical power. At first they were hard to obtain; the ships were small and sharks perhaps dangerous. Moreover, the magical powers sought had a strange history, being derived from a goddess who had originated through some violence done to a god, a god who through this hurt was a benefactor of mankind. It is hard to avoid a suspicion that the color of coral, and perhaps of Tyrian

15. Elliot Smith, *in* Jackson, *Shells as Evidence of the Migrations of Early Culture.*

purple also, implies his blood. This remote and uncertain history has had an enormous influence on the creative ability of man. Traces of it survive in the folklore and in ornaments of many people. As the original stories became more remote in time they became more majestic and have influenced the greatest poets right up to our own time. There are many things that are obscure in this history, and some of them can only be illuminated by further studies of the organisms that carry the myth. The distribution of practices must be more closely compared with the distribution of such organisms. Biological oceanography has some part to play in the final elucidation of the histories of coral, cowries, Tyrian purple, and pearls. Our science may have wide peripheral meanings and may help to light up very deep recesses of the human soul. And so, with Prospero:

> Now my charms are all o'erthrown
> And what strength I have 's mine own
> Which is most faint: now, 'tis true,
> I must be here confin'd by you
> Or sent to Naples.
>
> . . . .
>
> Gentle breath of yours my sails
> Must fill or else my project fails,
> Which was to please. Now I want
> Spirits to enforce, art to enchant
> And my ending is despair
> Unless I be reliev'd by prayer
> Which pierces so that it assaults
> Mercy itself and frees all faults
> As you from crimes would pardon'd be,
> Let your indulgence set me free.

PASCÀ: A NOTE ON THE VENERATION OF SOULS IN PURGATORY

After mass on Monday mornings during the 1920s, a group of women collected in the chapel on the right of the high altar in the church of San Giovanni a Mare, near the water front of Naples. The church is a Gothic building, once associated with the Hospitalers of Jerusalem, later known as the Knights of Malta. It is now sadly ruined, with its entire interior destroyed, but then it was filled with elaborate ecclesiastical furniture. A viaticum coach stood behind the high altar, the eye of God depicted on its back window, surveying the proceedings in the church. At the western end of the building, a glass case was filled with figures for the Christmas *presepio*, angels of baroque weightlessness hanging on wires like hummingbirds in a Victorian natural history museum. To the fifteen or twenty women who had gathered in the church, what mattered, however, was a skull in a hole in the pavement.

Every Monday morning, a stone in the floor in front of a shrine between the altar of the chapel and the high altar was removed, disclosing a pile of skulls with a few long bones. Only one of the skulls was of any importance, one with an unusually perfect set of teeth, placed on top of the others and ornamented with a wreath of artificial roses. This skull was said to be that of someone called Pascà (Pasquale) who, though in Purgatory, was believed to be somewhat closer to the Blessed Virgin and her Son than are those of us upon earth. In return for masses and prayers said for his repose, it was believed that he could make effective supplication for his devotees standing round the hole.

Some Catholic theologians have affirmed, and others have denied, the possibility that souls in Purgatory know of the necessities and dangers of those on earth and can pray for them. In general, the belief that this is possible seems to have been emphasized officially mainly in the Counter Reformation and later. It may well have been a popular belief when Saint Thomas Aquinas declared it to be unreasonable. In the fifteenth century, Saint Catharine of Bologna, a most holy lady to whose arms the Blessed Virgin in a vision once entrusted her baby Son, is said to have had recourse to the souls in Purgatory whenever she desired any favor, her requests being immediately granted.

In the form in which the cult appeared in Naples, it probably never received any ecclesiastical approval. It was tacitly assumed that such things happened and could be tolerated, at least for a time, just as it was assumed that Neapolitan women habitually entered churches with uncovered heads. According to Alexander Polovtsoff, in *The Call of the Siren*,[1] the authorities finally closed up the hole, presumably in the 1930s. The accounts given by Polovtsoff, and by Cecil Headlam in *The Story of Naples*,[2] leave something to be desired. Headlam, as did the present writer, visited Pascà in the company of Miss Florence Allen, a lady of immense Neapolitan erudition, but he makes a glaring error about the Knights of Malta that would lead the unwary to suppose the church dedicated to the Evangelist and not the Baptist. These seem to be the only published references to Pascà. Neither writer mentions him by name.

The proceedings in the church of San Giovanni a Mare were led by a bareheaded middle-aged woman, in a black skirt, white blouse, pink knitted coat, holding a rosary. She stood on the south side of the hole, apart from the other devotees. At intervals as the rosary was recited, the leader and some of the other women engaged in conversation with Pascà, moving their hands slightly and talking as one might to a rather refractory child. Pascà was asked to intervene

1. London, Selwyn and Blount, 1939.
2. London, J. M. Dent, 1927.

in the various family matters that were discussed freely in the little crowd. Sometimes there would be scolding and shouting, as the numerous candles cast strange lights and shadows on the animated faces of the women in the dark church.

No one seemed to know when the cult started or who Pascà was, though it was said that the skulls had been discovered in or near the church by workmen. Some of the women who took part claimed that Pascà visited their houses, doubtless in dreams. A part of the belief on which such a cult might be sustained is easy to understand. If *de mortuis nihil nisi bonum*, a proverb inspired as much by a fear of retaliatory ghosts as by decent feelings, then these selfsame dead would automatically acquire an aura of respectability even if one did not know who they were. The belief that the Church Patient, though in Purgatory, has somewhat more direct access to the inhabitants of Heaven than does the Church Militant on earth, has, as has been indicated, some theological standing, and is no doubt reinforced by popular iconography. Every Monday, the day of the week dedicated to the patient souls, plaster representations of Purgatory were placed outside houses, each with a money box below it, as a means of soliciting alms to pay for masses for dead relatives. In these representations, a number of figures emerging from a sea of flames are supplicating the Virgin who in turn is appealing on behalf of the tortured souls to her Son on the cross at the top of the scene. If Pascà was thought of as one of these patient souls, the possibility of his slipping in a few practical requests on behalf of his earthly benefactors is by no means unreasonable.

It is said that this type of cult arose from time to time in Naples when skeletons were discovered, whose disincarnate owners might well be supposed to have no one to pray for them, and that Pascà merely provided a particularly well-developed example of such a practice. If it were not for a rather startling analogy, his soul could be left undisturbed, with such an explanation, whatever havoc the misfortunes of war have wreaked on his poor skull. *Requiem aeternam dona ei, et lux perpetua . . .*

On the eastern edge of Palermo, near the River Oreto and not far from the bridge built in 1113 by the Admiral Giorgios Antiochenos, is a small church dating from the late eighteenth century, though redecorated later, and dedicated to Santa Maria del Carmelo. It has at times been known as the Madonna del Fiume or del Ponte, but is more usually spoken of as the Chiesa dei Decollati. The naming of the nearby Piazza dei Decollati on a street map may, in fact, be the first intimation to the visitor that there is something strange about the region, but the church is of no artistic interest and most travelers never visit it.

Within the church, on the right of the nave, there is a chapel dedicated to Saint John the Baptist; above the altar there is a painting of the decapitation of the saint; below the altar slab is a large *Purgatorio* of the kind already described. Here women come to pray after the Monday morning mass.

The cult has been described in detail by the great Sicilian folklorist Giuseppe Pitrè.[3] It is of some importance; in his time, Pitrè could ask who among the women of Palermo had not taken part in it. Though perhaps it is now practiced more quietly than formerly, its essential features remain unchanged. In one respect there may be a slight alteration in emphasis and interpretation. Pitrè says that the souls of the decapitated are spoken of as *armi di li corpi decullati* and are so distinguished from the *armi santi*, the ordinary term for souls in purgatory. Today the notice directing visitors to the church uses the expression *anime sante dei corpi decollati* and perhaps in general the distinction is now blurred, though even when Pitrè wrote, he says it was not always made.

The souls in question are those of executed criminals, whose bodies were buried and heads formerly exposed in the neighborhood. In the eighteenth century, the skulls of the exposed heads

3. *Usi e Costumi, Credenze e Pregiudizi del Populo Siciliano*, Vol. 4 (Palermo, 1889), reissued as *Opere Complete di Giuseppe Pitrè*, Vol. 17 (Florence, G. Barbèra, 1952).

were apparently finally buried in or near the church. To those who had faith in them, the *armi di li corpi decullati* were of great assistance on earth. Pitrè says that they helped fight the Bourbon army in 1860. Whatever their behavior may have been in the flesh, they are, after death, on the side of the innocent and defenseless. They were evoked by victims assaulted by thieves, and then might leap from their graves as skeletons, blowing trumpets, and seizing odd femora or humeri from the cemetery as weapons, put the malefactors to flight. Such scenes were depicted in popular prints and votive paintings. Under more normal circumstances, they were frequently consulted as beings who could grant requests and make known the outcome of future events.

The full ritual as given by Pitrè took the following form. The devotee, who seems ordinarily to have been a woman, set out early on Monday or Wednesday morning to the church, removing her shoes for the last part of the journey. The initial prayers were said on this last part, consisting of a rosary of one *Paternoster*, three *requiem aeternam*, and one of several *misteri* appropriate to the occasion. Then came the *Ave Maria* ten times, ending with a prayer to the Savior recalling his flagellation and asking him to comfort the souls of the decapitated.

Several of the *misteri* employed are given by Pitrè. In one version the souls are said to be born on Earth and now to be in Purgatory, waiting for Paradise; they are asked to pray to the Eternal Father for the needs of the devotee. From other prayers, we learn that the souls are nine in number, three of men hanged, three of men killed, and three of men drowned. The hanged and drowned men were no doubt decapitated post-mortem; hanging was usual in Palermo for the lower classes of criminals, and beheading for the more distinguished malefactors, but drowning seems an odd mode of execution. The nine souls enumerated in such prayers seem also a curious underestimate of the number of criminals executed in Palermo who would still benefit by the prayers of those on earth. Though the interpretation is far from clear, the phrases

Tri 'mpisi, tri ocisi e tri annigati
Tutti novi vi junciti[4]

suggest that there is something more to the *armi di li corpi decullati* than is obvious from the cult as it is now known. The conjunction of drowned and hanged men has curious overtones, suggesting *The Tempest* and *The Wasteland.*

The full set of prayers was repeated fifteen times, and the Lauretian litany said, by which time the devotee was at the door of the church. Once inside, the specific request was made at the altar of Saint John the Baptist. The devotee then moved to a stone to the right of the chapel, addressed the souls supposedly beneath it, and put her ear to it, listening for a light tinkling sound from the *armi di li corpi decullati*, signifying that her desire would be granted.

Invocation was also made at home, at night. A candle would be lighted in front of a representation of purgatory. The door of the house was then set ajar or a window or shutter just opened. Nowadays it is said that the invocation should be from a balcony rather than in the house. The devotee, on her knees, began with reciting the *Gloria Patri*, and then five *poste*, the same sequences of prayers that were used in approaching the church. The request was then clearly announced, and a final prayer threatened the souls with neglect if the devotee was not granted what she had in her mind. This was repeated on nine successive nights. During the saying of all the prayers the devotee also listened for sounds breaking the silence of the night, which might be interpreted as the echo of the souls.

The sounds that come through a Sicilian window even now seem always to carry emotion; a donkey cart passing or high heels on a

4. *Tre appiceati, tre uccisi e tre annegati*
Tutti nove vi aggiungete

'Three hung, three killed and three drowned
In all you make nine.'

Ocisi = *uccisi* here doubtless means decapitated. Pitrè gives this prayer as used during invocation at home.

pavement are very different from a distant Vespa or loud-speaker even though the intensity of the sound may be the same. Pitrè gives a list of favorable and unfavorable sounds which indicated whether or not the request to the *armi di li corpi decullati* would be granted. It was good to hear the crowing of a cock, the barking of a dog, a good whistle, the sound of a guitar, a voice singing well of love, a peal of bells, a loud ringing or knocking at a door, the closing of a shutter, the rapid passing of a carriage. The bad signs were caterwauling, the braying of an ass, a quarrel, lament or plaint, a sudden wailing of wind, or some water thrown into the street. The last sign was the worst. Words also might be heard, giving auguries in oracular form. Similar sounds were listened for at the feast of Saint John the Baptist and on other occasions.

The cult of the *armi di li corpi decullati* was not confined to Palermo, but naturally centered there as the chief seat of the administration of justice. It achieved a certain degree of official sanction. Pitrè quotes from a *Sermone per le anime de' Decollati* by the Rev. Fortunato Mondello, recalling the penitent thief and assuring his listeners that the gallows had been purified by the cross. Some of his passages sound a little like Simone Weil on civil punishment as a religious rite:

> *. . . et latronem exaudisti*
> *Mihi qnoque spem dedisti*

Pascà was in Purgatory, but could be supplicated; he lay under a stone to the right of an altar dedicated to Saint John the Baptist; he was, in the state in which he was visible, most certainly *decollato*. In these three ways he resembles the *armi di li corpi decullati*. It is just possible that the proximity of San Giovanni a Mare and the Chiesa dei Decollati to water might have some significance. The mode of veneration in both places was clearly reminiscent. The Sicilian cult is more violent, yet more poetic and immaterial: it explores the relationship of the New Testament to the Law in an imaginative but fundamentally Christian way, which would appeal strongly to a people who have experienced all but continuous alien rule throughout their history. The Neapolitan cult is more localized, more matter of fact and raises fewer problems in moral theology.

In neither case are there any documentary or archaeological evidences of antiquity. Pitrè points out that in the Middle Ages the souls of condemned criminals were supposed to harass any priest who was bold enough to offer them consolation before death. He seems to link the cult of the *armi di li corpi decullati* with the institution of the *Compagnia dei Bianchi* and other charitable organizations set up in Sicily in the sixteenth century to provide spiritual comfort to condemned criminals. The *Bianchi* played an important part in the religious life of Palermo. It would be possible to interpret both the cults as quite independent developments. Yet in view of the curious similarities this does not seem very likely. Alternatively one could speculate wildly, inventing a ninefold human sacrifice to a water god now assimilated to Saint John the Baptist, once widespread and now surviving in different vestigial forms in Naples and Sicily. For this there is no evidence. We can look into the dark, see nothing and hear only voices that cannot be interpreted. This does not mean, however, that there is no history behind the meager written record, but merely that there is now no means of knowing what the history was. Only recognition of some similarity in the writings of antiquity or the bringing to light of another existing cult, comparable to Pascà and to the *armi de li corpi decullati* but with an additional illuminating feature, could now tell us what these mysteries really mean.

SIX DISSOLVING VIEWS

1. The Barbarians

Professor J. G. D. Clark has written a book[1] about the Barbarians, the people who hacked their way into the deciduous forests of Europe, and established hunting and farming societies which lay between the great urban cultures of Egypt, Mesopotamia, Phoenicia, Crete, and, later, Greece and Rome, and the savage Hyperborean tribes who shivered in the coniferous forest and tundra of the far North. The Barbarians, who were the ancestors of the present writer and many of his readers, laid the material foundation of modern Western Europe and so, indirectly, of contemporary North America. This they did by clearing large area of the forests of Central Europe westward to Britain, using, in the Neolithic period, stone axes. On the cleared land, peasant farming communities developed, trade was possible with the more civilized ancients around the Mediterranean, and a variety of technological processes were introduced or developed. The economics of such people inevitably were economics of subsistence and of barter, so that Clark's book is as much an ecological as an economic study.

Perhaps the parts most fascinating to the natural scientist concern the introduction of a host of new techniques into archaeological research. Nearly everything has disappeared from even the best prehistoric site; the art of the archaeologist is to make the little which

1. *Prehistoric Europe: The Economic Basis* (London, Methuen, 1952).

remains yield the maximum information. The value of pollen analysis in providing a relative time scale for the European prehistorian has been recognized, at least in Scandinavia, for a quarter of a century or more. It is illuminating to see how much information can be extracted from the pollen profile of a bog if it is skillfully interpreted. Clark gives from Iversen a fascinating diagram of the section through Ordrup Mose in Denmark. In the middle of the profile, charcoal appears in quantity. Above this, birch increases and oak declines. The total tree pollen exhibits a pronounced minimum at the charcoal layer and in all succeeding levels the pollen of cultivated cereals and of weeds such as plantain (*Plantago*) and dock (*Rumex*) are present in abundance. This bog records the introduction, with the first Neolithic settlers of the region, of the *Brandwirtschaft* or agricultural clearing by burning off forest, which reverts after a few years to birchwood; this has till recently been a major practice in Northern Europe.

Another less familiar technique rests on the carelessness of the Neolithic and Bronze Age housewives. Pots were made at home, and, if the floors on which they were placed to dry had not been properly swept, the bottom of the pot was liable to acquire the impressions of whatever had been scattered accidentally on the floor. By studying a sufficiently large series of potsherds from any given horizon, it is often possible to identify the commonest objects which escaped the broom, such as the seeds of edible cereals. In this way, the relative abundance of wheat, barley, or millet can be determined, and the changes in the crop with changing culture and changing climate ascertained. Wheat was more important than barley in both Denmark and Britain during the Neolithic, while barley all but displaced wheat in the Bronze Age, though it returned to equal importance with wheat in Britain, but not in Denmark, before the Roman invasion of the former country. Oats and rye appear late; their growing importance in the Iron Age is no doubt due to their tolerance of the climatic deterioration then taking place.

It is often possible, from a study either of the bones of migratory wild animals or from the distribution of age groups in animals either wild or domestic, to ascertain if a site were occupied during the whole year or seasonally. It is evident that the Hamburgian culture of the Upper Palaeolithic of northern Germany represents a group who moved northward to hunt reindeer in the summer. In winter they may have lived much farther south, perhaps in caves, and may conceivably be represented by quite other names in the archaeological literature.

Petrologic study, when applied to the Neolithic stone axes in Britain, reveals cases of extraordinary localization of the sources of raw materials. It was evidently as important to the Neolithic man of the south coast of England to swing a Langdale or a Graig Lwyd axe, as it was to a later Englishman to have a Sheffield knife in his pocket. A few exotic persons even seem to have insisted on, or at least to have acquired, axes from Tievebulliagh in Northern Ireland. Metallurgy, boatbuilding, and house architecture yield similar fascinating glimpses into barbarian life. If at times the poetry and the feeling of the hand on the material seem to be missing, it must be remembered that the best work of these peoples probably was given to textiles and to wood.

The inevitable, almost complete absence of reference in Clark's book to ceremonial life or to beliefs may have obscured certain aspects of his subject. Some explanation is surely needed of the magical significance of iron, when it appears that, of all metals used in pre-Roman Europe, iron was the only one which could be, and was, made the basis of a nonprofessional home industry. But, whether we like it or not, these people were perhaps practical rather than philosophic, and it may be that the fifty-seven wine strainers, obviously imported at great expense from the Mediterranean countries, which have been recovered from various Danish sites embody their highest aspirations and form their most fitting monument. Whatever their philosophic attainments may have been, we owe much to the Barbarians and much to Professor Clark for ordering the vast amount of information which is becoming available about them.

We may reasonably guess at a Neolithic origin for just one far-reaching if erroneous idea that played a considerable part in early science. The seasons, the temperaments, and the qualities of hot and cold, wet and dry have fascinated philosophers and artists. That the coldness or hotness, dryness or wetness of matter should seem so fundamental may well be an attitude which grew up among the women who first made pottery. Nothing done before, perhaps, could have given so great a feeling of creation as that of making a pot. Later, when the potter had become a man, he easily developed into the prototype of an omnipotent God.

II. The beginning of wisdom

There are countless histories of philosophy and an ever-growing number of histories of science. In such works it is usual to trace out the origins of ideas which have proved fertile and thus to view the ancients in terms of contemporary preoccupations. Of the recent scholars who have attempted an unbiased approach to the problems of intellectual history, none has been more successful than the late F. M. Cornford. He started his career investigating the subject and he returned to it many times, embodying his conclusions in a book all but complete at the time of his death. This book, *Principium Sapientiae*,[2] bears on the origins of practically the whole of our mental life and is of immense interest to students of the history of science, no less than of philosophy, poetry, and religion.

In the *Phaedo*, Socrates is reported to have said that, in his youth, he was an enthusiast for what was called *περὶ φύσεως ἱστορία*, usually translated as natural science, but that he became dissatisfied with such studies. This has made him a sort of patron saint of the departments of the humanities in universities that find satisfaction in the thought of so eminent a philosopher rejecting physics in favor

2. F. M. Cornford, *Principium Sapientiae: The Origins of Greek Philosophical Thought* (Cambridge Univ. Press, 1952).

of a belief that the proper study of mankind is man. It is, of course, obvious that Socrates does mark a break in Greek thought. It is also obvious that, within two philosophical generations, Aristotle's biology appeared as one of the first large bodies of work in pure science in Greece that really amounted to anything. Aristotle in his other writings on ethics and literary criticism shows the new direction imparted by Socrates, but his biology has quite different roots. It is, however, legitimate to ask if these roots lay in the kind of ἱστορία that Socrates had rejected. Cornford answers emphatically, No.

Cornford begins by examining the atomists, Lucretius, Epicurus, and Democritus, and the earlier Ionians behind them, and asks whether they were scientists doing experiments. Again Cornford's answer, given perhaps a little too emphatically, is No. He then quotes from an Hippocratic writing, *Concerning Ancient Medicine*, a warning that considerations derived from natural philosophy should be excluded from the empirical study of medicine. The natural philosophers discoursed on things about which no information could be obtained. The passage reads just like a modern paragraph by a behaviorist, or an operationally minded physicist, or a logical positivist, protesting the importation of metaphysical nonsense and other nonobservables into science. Moreover, the later atomists, Epicurus and Lucretius, in their desire to purge the universe of the irrational, were as dogmatic as any of their Neoplatonic antagonists.

What the Ionian philosophers were trying to do, according to Cornford, was not to produce empirically testable hypotheses of cosmogony, but to rationalize existing cosmogonic myths. Such a myth is preserved practically intact in Hesiod's *Theogony*, and can be related (and thanks to the *Epic of Kumarbi*, as E. R. Dodds points out in a footnote to Cornford's text, related better than Cornford knew) to Mesopotamian myths of the same kind. These in their turn are now usually believed to be rationalizations of rituals and to refer initially not to the inception of the universe, but merely to that of the agricultural year.

The true origin of Western science lay not with the Ionian philosophers but with the empirical natural histories of disease worked out by Hippocrates and his successors. When a pure biology begins to take shape, it is not surprising to find that it is the work of a man of wide curiosity brought up in a medical family. This biological science, in so far as it is valid, is, however, mainly descriptive.

The initiation of natural history is not the only Greek contribution to modern science. Some scientists, not being naturalists, would in fact say that it was of no great importance. The other great advance was the introduction of mathematical generalization. Here again the history is complicated. Cornford thinks that the prototype of the Greek intellectual resembled the Siberian *shaman*. The aspect of the shaman, or the divinely inspired poet, that Cornford emphasizes is his knowledge of the hidden things of the past, present, and future. Cornford quotes several ancient authors, both Latin and Greek, to the effect that the divine inspiration of Apollo conferred a vision not merely of the future but disclosed *quod eritque fuitque estque* (Ovid, *Metamorph.* I.517), much as in the Christian doxology, "as it was in the beginning, is now, and ever shall be." In historic times, the activities of the shaman become dissociated into those of the seer, the poet, and the philosopher, and are by no means mutually consistent. In Plato and his Pythagorean predecessors, shamanistic timeless knowledge, the ability to see past, present, and future, still lingers. It appears as the eternal truths of mathematics, and in our ability to recognize such truths, because we are in a sense timeless. This recognition of the special place of mathematical reasoning is fundamental in the whole post-Platonic history of science. The mathematics may already have been known; even the theorem of Pythagoras states a result which was available at least empirically at a much earlier date in Mesopotamia. What was important was the realization of mathematical form as an enduring skeleton by which our knowledge of the external world can be supported and systematized. It does not matter historically that the Platonic attitude to mathematics is at the moment out of fashion.

III. *Ky bien pense bien puet dire*

No period of Western history suffers from a greater amount of misunderstanding than the High Middle Ages. The very term "medieval" is frequently used as a term of abuse; labor conditions on the New York water front were recently described in a responsible newspaper as medieval, meaning, apparently, early nineteenth century. A good many working scientists probably still believe that their subject was virtually nonexistent during the whole period.[3] It is often held that the main reason for this supposed dormition was the intellectual tyranny of the Western Church, broken only at the Renaissance and the Reformation. Actually, these ideas are now appearing as false. Everyone must admit that a vast amount of superstition and misinformation was current during the period; it is perhaps harder to admit that successful twentieth-century astrologers were, until artificial satellites acquired a political significance, probably able to earn a better living than the most eminent astronomers. It is also obvious that, in most of Europe during the sixth, seventh, and eighth centuries, things were not what they had been. But somewhere in the countries around the Mediterranean there was always an interest in the natural world and in what had previously been written on the matter. Moreover, in the long run, the sacramental nature of Christian worship and the Christian emphasis on charity, which involves operation in the external world, as the supreme virtue have been far more favorable to the rise of science than asceticism and ecclesiastical intolerance have been unfavorable. Some recent writers[4] have expressed a belief that Western science is inevitably bound to Christianity in the future as in the past; certain recent developments would seem reasonably interpreted in such terms.

3. A. C. Crombie, *Augustine to Galileo: The History of Science A.D. 400–1650* (London, Falcon Educational Books, xv, 1950).
4. For this point of view see, among other works, John Baillie, *Natural Science and the Spiritual Life, being the Philosophical Discourse delivered before the British Association for the Advancement of Science at Edinburgh on 12 August 1951* (Oxford Univ. Press, 1951).

The Renaissance at first was unfavorable to science through its excessive preoccupation with ancient literature; and Protestantism, though officially a movement of liberation from superstition, quickly developed an irrational bibliolatry which was later a far greater barrier to scientific development than the ritual sacramentalism of the Middle Ages.

The circumstances of the rise of natural science in the Western world are complicated and curiously contradictory. There was first of all a technological basis which continued to develop steadily from the Dark Ages onward. The fact that a considerable part of this technology was devoted to the production of ecclesiastical architecture on an increasingly vast scale, and another part to the problems of warfare and the aggrandizement of princes, has undoubtedly led to a general undervaluation of its intellectual significance. The mechanical problems of enclosing space with a vault are the same whether the building is "for a scanty band of white-robed Scholars only" or for a sea of variously garbed commuters. If the thing can be done, the knowledge to do it must be available. Crombie, in a long chapter on "Technics and Science in the Middle Ages," indicates in instance after instance how the technical knowledge of the time was continually growing and interacting with the theoretical interpretation of the external world. This interaction was in fact recognized in educational practice.

The second requirement in setting the stage was a very curious philosophical situation, but one which keeps on re-establishing itself today. The great scientific period of the Middle Ages began in the thirteenth century. Historically, the activity of this period drew its past from philosophic speculations in various places. One of these was Chartres, where the outlook was Augustinian and Platonic. The Platonic interpretation of reality, in terms of eternal forms that could be conceived of as exhibiting true relations of a mathematical kind, gave mathematics an intellectual primacy at a time when practically no mathematics, as we know it, was known. "Seven and three are ten and not only now but always; nor have seven and three in any way at any time not been ten, nor will seven and three at any time not be ten" (Augustine, *De Libero Arbitrio* II. 8. 21).

At the same time that Thierry, Gilbert de la Porrée, and Bernard Silvester were emphasizing Augustinian ideas of such Platonic generality at Chartres, nominalism, first propounded by Roscelinus, who was at Chartres for a time, was beginning to be important in a less extreme conceptualist form as the result of the activities of Abelard in Paris. A completely realist (in the medieval sense) science would be impossible because, if the eternal ideas alone had reality, the individuals with which science deals would have no distinguishable criteria—unless, of course, one postulates an eternal idea in one-to-one correspondence with every individual difference, which has been done, but which renders the ideas pointless and unnecessary. A completely nominalistic science would seem to be virtually impossible psychologically, in spite of recent valiant efforts to construct it. It is extremely difficult to conceive of the sentences stating scientific generalizations as referring to nothing between the discrete identifications of the propositional variables as individual names. One may know that the sentences are really recipes for their own confirmation, but one wants them to have something like mathematical existence, whatever that may be, between the moments of actual use in specific situations. It is, therefore, probably no accident that the rise of nominalism and its head-on collision with a well-developed Platonic realism were antecedent to the thirteenth-century rise of science.

About the same time, Aristotle began to become available in translations, in part from the Greek, in part from the Arabic. During the succeeding century, the literature of science came to mean Aristotle. This was inevitable, since the man had in fact done so much. It was also partly in keeping with the nominalist criticism of the Augustinian position, which was in a way unfortunate because it led to bad physical explanations being substituted for better mathematical descriptions. The very fortunate fact that Aristotle lived three and a half centuries before Christ prevented his being taken as the ultimate authority on everything, by everyone except for a few Averroists, who were condemned by the Archbishop of Paris in 1277. The two outstanding Aristotelians, Albertus Magnus, whom we may honor for many things including the discovery of the typical life-history of the higher insects, and Thomas Aquinas himself, were both men of great common sense.

This very brief summary of a theme developed, though a little differently, by Crombie in the first two chapters of his *Augustine to Galileo*, will at least give some idea of the intellectual excitement of the thirteenth and early fourteenth centuries. The achievements were also remarkable. Crombie devotes just under a hundred pages to a summary of the state of science during the period. The work is, of course, from a modern point of view, nearly all preliminary. Practically no major generalizations that have survived till today were made during the period. Some of the individual achievements, notably accurate determinations of latitude and of the obliquity of the ecliptic, are impressive; while of greater general importance was such work as that on the foundation of statics[5] by Jordanus Nemorarius in the early decades of the century, and of optics by Grosseteste, Roger Bacon, Witelo, and Theodoric of Freiberg, which provided bases on which all subsequent developments were grounded. Throughout this time, new knowledge of Arabic science and Hindu mathematics was reaching Europe; and the ordinary Arabic, really Hindu, numerals were becoming widely available to the Western reader in the work of Leonardo Fibonacci of Pisa, a mathematician of considerable significance for other reasons.

This was a magnificent beginning. That it was no more than a beginning is partly due to the inherent difficulty of making the

5. A number of important medieval works on this subject have been made available in the original Latin and in English by Ernest A. Moody and Marshall Clagett, *The Medieval Science of Weights, (Scientia de Ponderibus): Treatises Ascribed to Euclid, Archimedes, Thabit ibn Qurra, Jordanus de Nemore and Blasius of Parma* (Madison, Univ. of Wisc. Press, 1952). This work, with much introductory matter and critical apparatus, is a notable contribution to the history of medieval physics.

simplest discovery, a difficulty well known personally to all scientific readers of these remarks. It is also to be noted that the thirteenth and early fourteenth centuries were politically and socially better times in which to work than the late fourteenth and fifteenth centuries. Part of the slowing of the pace noticeable in the late fourteenth and early fifteenth centuries is certainly due to social and economic disturbance. Part is no doubt also due to the tremendous literary impact of the High Renaissance. As early as 1200, John of Salisbury was lamenting the current trend away from the humanities, just as if he had been dean of a graduate school in twentieth-century America. The trend was reversed with a vengeance by 1450.

One further reason that the initial enthusiasm for science could not be maintained may have been a shift in the philosophic climate. Nominalism continued to develop and finally reached a quite Humean position with Nicholas of Autrecourt (second half of the fourteenth century). The work at Oxford of William of Ockham (1300–49) produced an essentially modern theory of induction. But one feels that, with the improvement of the theory by gradual elimination of eternal universals, the interest of the process in practice seems to wane. By the time of the opening of the great scientific era of the late sixteenth and seventeenth centuries, some of the old mathematical realism, derived from the literary interest in Plato, had crept back to reinforce a physicalism of ultimately Aristotelian origin. This in fact was the fundamental intellectual source of Galileo's difficulties with the Church. At least the more liberal churchmen would have had no objection to the use of the Copernican system as a mode of calculation. But Galileo insisted on an absolute rather than a relative heliocentrism. Three hundred years later, Einstein demonstrated the position for which Galileo was imprisoned to be meaningless.

IV. Apes and *Silvatici*

A curious sidelight on medieval science has been cast by the immensely learned work of H. W. Janson on *Apes and Ape Lore in the*

Middle Ages and the Renaissance.[6] This is a study, both iconological and literary, of the knowledge and understanding of subhuman primates in Europe from the beginning of our era to about 1700, with a few asides on literary pithecology in the eighteenth, nineteenth, and twentieth centuries. For the period and region which it covers, it is a Pithecopedia, though it claims not to be, a *singerie complète*. It is marred by a few neologisms and attempts to be smart, which will distress the fastidious reader, and by a persistent inability to capitalize italicized generic names, which will outrage the systematist. It is, however, an extraordinarily interesting book.

The ancients, according to Janson, were not distressed by the apes. They may have worshiped them, killed them, or dissected them, but they did so without searching of soul. During the earlier Christian era, the ape appears first as the devil; the evidence here is scanty and the story not entirely convincing. It is evidently full of Freudian undertones, but these are avoided as is customary in such works. Later, the ape appears as a symbol of sinful man, as an impostor aping human ways. He gets in the trees in the Garden of Eden and sets Eve a bad example by eating forbidden fruit. The whole story is long and very complicated; one myth comes as another goes. But throughout the whole period, whatever else the ape is doing, he is posing a problem, the problem of his relation to mankind. Aristotle and Galen think his internal structure is essentially human. Thomas of Cantimpré, though he can have had no experience in the matter, takes the opposite view. Bartolomaeus Anglicus agrees with Aristotle, Albertus Magnus follows Aristotle in one place, Thomas of Cantimpré in another.

Pygmies are continually intruding their presence. St. Augustine had taught that any living being which is rational is human. Albertus, who got his apes almost exclusively and his pygmies entirely from classical sources, considered that the apes and pygmies were to be classified as a special group of manlike animals, *similitudines hominis*. Within the group, the apes are supposed to control their

6. London, The Warburg Institute, 1952.

bodies solely in response to immediate sensory data, though with some degree of insight; the pygmies have memory and can learn from experience, but can form no generalizations. The power of such abstraction belongs only to man. Pygmies, however, can imitate human behavior, and apes, pygmaean behavior. In a certain sense, the ape with his obviously anthropoid structure and slight degree of insight, is a *resultatio obscura hominis*, a dim reflection of a man. The existence of such a reflection was evidently profoundly disturbing to Christian philosophers who proclaimed the infinite worth of every human soul.

Both in folklore and in the works of the learned, the existence of apes kept raising the question as to where they came from and what were they doing. Could they in fact be degenerate men? Christ and one of his disciples are entertained by a blacksmith; Christ puts the blacksmith's elderly wife into the fire and brings her out a strong girl of fifteen. Later, the blacksmith tries to repeat the experiment at the request of an old woman, who has to be rapidly rescued in a shriveled and blackened condition. This so frightens a couple of pregnant women standing by that they give birth to the first apes. Alternately, God visits Eve and asks how many children she has. Fearing to be censured for the results of overmuch lovemaking, she admits to only part of her family, concealing the rest. The unacknowledged children become apes.

Variants of these tales have an immense diffusion. A close parallel to the first is known from Annam. Janson dismisses the parallelism with a remark that "men everywhere tend to be 'Darwinist in reverse' when faced with the problem of explaining the existence of apes," but this hardly accounts for the rejuvenation of an old and ugly woman and the use of fire in the story. Somewhere, something must be hidden in the history of the myth, though admittedly its persistence and diffusion testify to its psychological importance. Janson is certainly correct in his belief that Darwin had to face not merely unprejudiced criticism of a purely scientific kind, but also a horrified suspicion that what had been taken to be a *resultatio obscura* might turn out to be a cousin if not an ancestor.

... the wild man went his weary way
To a strange and lonely pump.
—Lewis Carroll

A cognate theme, which in part overlaps that of the ape, has been explored by Richard Bernheimer in his *Wild Men in the Middle Ages.*[7] This book deals initially with men who have fallen from human stature, usually as the result of various kinds of emotional shock, and who populate the forests of the medieval romances. In their most sophisticated form they are often wellborn men who from misadventure in love have lost their reason and have taken to the woods, feeding like Nebuchadnezzar and the Prodigal Son on vegetable matter fit only for animals. They can, however, be won back to human society by force or by kindness. They thus become symbolic figures of the irrational energetic aspect of the human mind, of unconscious drives, of the Freudian id. In the tapestries and caskets of the late Middle Ages they appear, like the ape, captive and not infrequently led by a lady. They here play the part of the unicorn, that most splendid of male symbols, tamed by the civilizing influence of a pure and beautiful woman. This theme is so insistent, whatever iconographic form it may take, that it clearly represents one of the main psychological preoccupations of the time, and loses none of its importance from being so distant from recorded daily practice. It is, however, only one aspect of the wild man.

The wildness of the wild man is not only a reversible lapse from humanity; he also had an independent existence. He could marry a wild woman and beget wild children. When he came in contact with ordinary men he might be killed rather than rehabilitated, and then, at least on occasion, seems to have been capable of a fleeting

7. *Wild Men in the Middle Ages: A Study in Art, Sentiment and Demonology* (Cambridge, Harvard Univ. Press, 1952). It is rather unfortunate that Janson's and Bernheimer's works should have appeared simultaneously as they are neatly complementary in matter and one would have liked to have known what one author felt about the other's material.

resurrection. It is such an autonomous woodhouse or *silvaticus* that still tempts the thirsty to enjoy a glass of beer at the Green Man in any of a great number of English villages. The heads of such *silvatici* are freely represented, often with branches growing from their sides, mouths, or even noses in English medieval sculpture. In some churches grotesque heads are more numerous in the north aisle than the south (e.g. Banwell and Winscombe in Somerset). Such a tendency hints at a pre-Christian origin.

Bernheimer clearly recognizes that behind the symbolic wild man on the tapestries and ivory cosmetic boxes, there lurks a more primitive crowd of beings, clad in skins or covered with leaves and lichens. Such beings may be iconographically almost indistinguishable from our mythological ancestors Adam and Eve. Some representations of the wild man take on the rude facies of various primitive peoples that were becoming known in the late Middle Ages. In this form at least the footprints of the woodhouse have survived into modern times, and have appeared in the press as those of the Abominable Snowman of the Himalayas, which some have supposed to be langur and some a far more remarkable being. However, the man of the woods is not always a rude primitive; in another form he appears as an inhabitant of the Golden Age, or as one of the pure and godfearing philosophers of India whose existence troubled the orthodoxy of the Middle Ages in the spurious Thoreauesque *Correspondence of Dindymus with Alexander*. He also appears as a fundamentally supernatural being who took part in various popular rituals of obvious pre-Christian origin. A man, woman, and child, clad as *silvatici*, were captured in the woods of the upper Inn valley and led in front of a great log on a carriage from tavern to tavern. This obvious fertility ritual was regarded as a sort of marriage, only to be celebrated in years in which no real marriage had occurred at carnival time. In other rituals the wild man was killed when caught; Peter Breughel the Elder depicts him being shot by a crossbowman, in front of the Emperor himself, while a woman in curious costume appears to be luring him by holding out to him a ring. Here marriage and death seem to occur simultaneously. This scene certainly depicts some village ritual, because there are attendant mummers taking up a collection from onlookers at the windows of the houses in the background.

At other times a whole Wild Horde may appear in such rituals; their behavior must underlie the antics of the Friars of Medmenham and similar aristocratic improprieties of the eighteenth century. Bernheimer traces a clear connection between the Wild Horde and the charivari or boisterous disturbance of the newly married couple on their wedding night, which is traditional in many parts of Europe and which has parallels in many other parts of the world. He traces some of the names applied to ritual wild men to Roman sources. A Spanish penitential of the tenth century conveniently proscribes one year of penance to those sinners *qui in saltatione femineum habitum gestiunt et monstrose se fingunt et maiam et orcum et pelam* ("who wear female dress in their dances and monstrously impersonate Maia, Orcus and Pela"). Who Pela was is not known, but Orcus, a god of the underworld, is still familiar in the word *ogre*, while the earth goddess Maia appears as *Holzmoia*, the word for a wild woman in some German glossaries. Bernheimer seems content to trace his theme to late classical times and to suppose that where there is a community of mythology between Eastern and Western Europe it may be attributed to a time before the decay of the Roman Empire. Here his caution is definitely misleading.

However, in another part of the work, where there is no question of trying to probe beyond history, we find the most extraordinary, though admittedly fascinating, speculations about a female figure wearing a dress with narrow shoulder straps (which must be unparalleled in medieval costume) and carrying a bucket and whisk that keeps appearing in manuscripts illuminated for King Wenceslaus of Bohemia. This girl, who is identified as a bath attendant, finally turns into that aspect of Sophia Euphemia, Queen of Bohemia, that could be paired off with the King as Wild Man. This psychological transformation is regarded by Bernheimer as a reconciliation of the old ideas of courtly love with the more bourgeois ideas of married respectability developing in the fifteenth century.

Thirty years ago every educated person knew that in the Arician Grove on Lake Nemi there was a sanctuary of Diana, whose priest was termed the King of the Wood and who was by any standard the wildest woodhouse of them all. A runaway slave who had killed his predecessor and who was fated to be slain by his successor, he was at the same time the ritual impersonation of a shadowy god Virbius, apparently the consort of Diana who here appears as a fertility goddess. He was also, as his name *rex nemorensis* implied, a sylvan deity apparently connected with the oak tree on which grew the Golden Bough. The King of the Wood in the Arician Grove takes the story well back into the Roman Republic and has every appearance of far greater antiquity. If we assume that where literary and iconographic evidence is lacking there never was any, we get a distorted picture. Actually, it is more reasonable to suppose, as the paleontologist would do, that all our information is fragmentary. The matter is of some human importance because if the wild man is the descendant of all the gods and all the earliest men of the West he is obviously more important than if he is merely some obscure classical deity diffused just in time to be conquered by Christianity.

These paragraphs were originally written because a king had died and because his successor, a queen, was shortly to be crowned, in a ceremonial that is partly a birth and partly a marriage. Thousands of people who have repudiated the political ideal of kingship availed themselves of the facilities provided by electronic engineering to follow this ritual, and very many paid much money to watch for a few moments the progress of the newly crowned queen. She entered a church as a womb, in the red cloak, the color of arterial blood, and left it in venous purple.[8] Her namesake and predecessor

8. The color symbolism I owe to Dr. Grey Walter, but was somewhat startled to see in the film of the Coronation how well it seemed to fit. It must be comparatively modern. The marriage of the sovereign and country is possibly involved in the crowning itself, and is comparable to the Doge of Venice marrying the sea.

prayed "that I may feed Thy people with a faithful and true heart."

The King of the Wood is not dead. To deny him would be as wise as to deny ourselves our temporal bone merely because *Australopithecus* possessed one. If we honor him rightly he will come as the unicorn or the sylvan philosopher or the phallic blackamoor that supported respectable city houses in the seventeenth century. If we deny him, or perhaps what is worse, if we regret him, he will come as the Wild Horde, and this we know, more than any generation for some centuries, can be very terrible.

v. What We Now Do in Science

The modern attitude has recently been examined and re-expressed in a remarkable book by R. B. Braithwaite, *Scientific Explanation*,[9] a work of the greatest importance to everyone who wants to understand what one does in science. Braithwaite, perhaps in a bicentennial mood, seems to have some of Bishop Berkeley's scruples about infinitesimals. He therefore founds the mathematical parts of his book exclusively on schoolboy arithmetic, algebra, and geometry, and what can be developed from these disciplines. The result is a quite difficult book, but difficulty would seem to characterize every honest attempt to reach primitive concepts.

Braithwaite starts his exposition by describing a scientific system as consisting of a hierarchy of hypotheses or general statements, the lower members being logically derivable from the higher. Some of the lowest members can be tested empirically, which means that particular instances of low-level hypotheses may be shown empirically to be untrue. If this is done, the system as it stands is known to be untrue, though in any well-developed science it is usually possible to alter part of the higher-level hypotheses so that the testable lower-level hypotheses are not found to be untrue. A valid system is one in which falsification has not up to the present been possible.

9. *Scientific Explanation: A Study of the Function of Theory, Probability and Law in Science* (Cambridge Univ. Press, 1953). Based on the Tarner Lectures, 1946.

Braithwaite in general accepts a Humean type of position in this matter. All his hypotheses, moreover, are statements about classes; only the particular instances which are tested empirically are statements about the individual members of certain classes.

The first step in the explication is the development of a calculus for a model theory. The calculus is initially a game, played by writing down certain formulae, which are arrangements of symbols or marks on paper. Then, by operating according to a set of rules, other formulae can be derived. Thus in Braithwaite's first calculus, the three initial formulae are

$$\alpha \longleftrightarrow (\lambda\mu)\ ,\ \beta \longleftrightarrow (\mu\nu)\ ,\ \gamma \longleftrightarrow (\nu\lambda)$$

Using five rules of substitution, which define the moves in the game, we can obtain, in seventeen operations:

$$(\alpha\beta) \longleftrightarrow (\gamma(\alpha\beta))$$

This calculus is so far completely uninterpreted. It can be interpreted in any number of ways. The way that is useful in the present context is to give the meaning to the symbol α: the class of everything which has the property A, etc. The double arrow is now given the meaning of class identity and the initial formula reads: everything which has the property A has the properties both L and M. We finally conclude that everything which is in both α and β is also in γ as well. Given the rules and the interpretation, this would appear to be intuitively obvious from the initial formula; it is an interesting commentary on our intuition that the formal proof involves seventeen steps.

It is now possible to consider a calculus, termed in Braithwaite's book S_3, which consists of an *algebraic* part in which the initial formulae contain variables, and a variable-free part in which the initial formulae are the same as those given above. The substitution rules remain the same. The algebraic part consists essentially of a system proposed by E.V. Huntington in 1933, and shown by him to represent the whole of the deductive system of class logic. Braithwaite's first major conclusion is that some such system as S_3 is needed in science, and that the variables of the algebraic part cannot be defined exclusively in terms of observables. If they are so defined, the algebraic part ceases to be algebraic and the theoretical system becomes rigid and is incapable of accommodating anything not known when it was initially conceived.

The inevitability of the theoretical terms, and of hypotheses expressed solely in such terms, raises the question of the "reality" of theoretical concepts. Braithwaite answers such questions as "Do electrons exist?" in two possible ways. The first way is to follow F. P. Ramsey and say that a property E, of being an electron, occurs in certain high-level hypotheses of a system, and further that these hypotheses are believed to be true since their logical consequences have not been found empirically false. In such a case, since the high-level hypotheses are taken as true, the theoretical properties mentioned in their statement may be regarded as existing in a certain sense. The second answer is that the property E is an interpretation of a symbol ϵ in a calculus, that ϵ plays a certain role in the calculus, and that the interpretation when consistently applied to the calculus yields low-level hypotheses, instances of which can be tested empirically and are found not to be false.

In either case, existence as ordinarily understood in this context means having a certain part in a system which is either interpreted or is interpretable, and which, after empirical study of instances of low-level hypotheses, has not been found to be false. It may be suspected that most scientists mean by existence in such a context that the concept persists in the deductive system down to a stage that is very close to the empirical test. Atoms and electrons would usually be considered as existing in a somewhat more concrete sense than the Schrödinger wave function.

To take an extreme case, the visual appearance of the cloud in a cloud chamber, of the scintillations on a zinc sulfide screen, of the tracks in a photographic emulsion, or of the clicks as a Geiger counter discharges, comes to fulfill the same role psychologically, bypassing the nature of the cloud, flash, track, or click in the minds of people who believe in electrons or atoms, as do the appropriate kinesthetic, tactile, and visual sensations in the minds of people who believe in tables and chairs. It is to be noted that a great many biologists who believed, without quite knowing what they were doing, in the existence of genes, have become somewhat disillusioned, because, although the evidence for their belief in the sense discussed above is perfectly valid, much modern information contradicts an over-particulate type of model, just as happened earlier in the history of the atom. Biologists, being less used to theoretical concepts than are physicists, have behaved less reasonably as their concepts have been changed and enriched.

In reading the earlier part of Braithwaite's book, one cannot help wondering what would happen if one tried to dispense with properties and classes altogether, as Woodger tells us we should do. It may be suspected, however, that radical as Woodger's ultranominalism seems to be, it does not differ from the position of the class logicians as much as at first appears. It may introduce chastity into an oversophisticated world, but it is legitimate to wonder whether in this case chastity is not merely clumsiness in disguise.

The greater part of Braithwaite's book is devoted to a consideration of probability statements in science. The problem is essentially as follows. At a high level in the system, an hypothesis H_p is introduced which states that the probability of the members of a class, defined by possessing the property β, also having the property α, is p. The question arises what we mean by this and how we proceed to an empirical test. Logically, a selection from the class, provided it was sufficiently large, can have any proportion of members exhibiting α or not-α. Whatever we encounter in an empirical case, therefore, can be *logically* consistent with the truth of H_p.

To Braithwaite, the hypothesis H_p corresponds to the proportion of members of a sufficiently large set that have the property α, this proportion being called the α-ratio. If we consider a die and say that the probability that it will fall with the 5 uppermost is 1/6, what we are usually saying is not that experiment shows a frequency of fives that is about 1/6 of the total number of throws, but that, having decided that the die is symmetrical we throw it in such a way that a great number of small forces operate on it. We can, therefore, make no prediction as to how it will fall, except one based on our knowledge that it looks like a homogenous cube with its faces numbered 1–6, but otherwise identical. Similarly, if we put a black and a white ball into a bag, having satisfied ourselves that one ball is not more slippery than the other and that they are of the same size, we may say that the probability of drawing a black ball is 0·5. This type of probability is sometimes called *logical probability*.

When we insert H_p into our system, it is such a probability which we are discussing. When we try to proceed to an empirical test, we can only determine frequency, or the α-ratio, in a limited set of actual circumstances. We know that any proportion is possible and require some sort of further rule of procedure to know whether we should reject H_p on the basis of the observed frequency.

Braithwaite takes as a model of a possible procedure n copies of the class defined by β, and having a proportion p also exhibiting α. Such copies can be visualized as black and white balls in bags, each bag having exactly the same proportion of the two kinds. A Briareus-like investigator with n hands now takes simultaneously one ball from each bag and in the array which he has collected determines the α-ratio r. Such an investigator is interested in what the probability is that he will be wrong if he rejected H_p on the basis of the α-ratio r in this observed selection of cases.

Braithwaite uses for illustration, since it involves no infinitesimals in its derivation, what he terms the k-rule of rejection, namely that if the ratio r falls outside the range $p \pm \sqrt{p(1-p)/nk}$, the probability that the investigator is wrong in rejecting H_p is less than k. Any similar rejection rule can be used and in practice most statistical procedures use a stronger rule. We have now replaced a logical probability by a frequency and the latter by something like a degree of (dis)belief. Braithwaite admits that as an account of probability in science this sort of presentation, which in his book occupies many pages and explores many ramifications in great detail, is unsatisfactory because it does not tell us what probability is. "It may well be said," he writes, that "it is circular to explain the original probability statement [i.e H_p] in terms of the circumstances under which it would be rejected, when rejection itself has to be explained in terms of probability." He considers, however, that, so long as we insist on rigorous explicit definitions, this difficulty is insurmountable because a rigorous explicit definition would be equivalent to translation of sentences containing probability words into sentences not containing such words, and this obviously cannot be done. The further explication must, therefore, not attempt explicit rigor. Braithwaite, however, points out that this is true of practically every basic concept; he has no patience with those philosophers who give up just when the problem gets interesting. In one special type of case, we can immediately go further. If $p=0$ or $p=1$, a single contrary case permits an absolute rejection of H_p as false. The connection between rejection and falsity can, therefore, be established at either end of the scale of probabilities.

The problem of the value of k to be taken in ordinary cases is considered. Logically, it is quite arbitrary. In practice, significance tests which do not require the rejection of hypotheses which have a probability of being wrong once in a hundred times are frequently used. Actually the value clearly depends on what we are doing with the hypothesis. A good deal of Braithwaite's book is devoted to a development of a theory of prudential strategy in selecting statistical hypotheses on the basis of empirical data. The theory is similar to that developed by Wald. The fundamental nature of such theories is that they take account, in describing scientific procedure, of certain evaluative processes which play a part over and above considerations of frequency in the rejection of an hypothesis. Most scientists are likely to be horrified by having to admit the validity of this approach, but it is quite easy to see in certain cases that something of the sort is happening. More unfavorable empirical evidence is required to cause rejection of an hypothesis that if true would be very useful, either practically or in subsequent development of a theoretical system, than would be needed if the hypothesis were relatively sterile. The hypothesis that there is a finite probability that a good chemotherapeutic agent in the treatment of cancer exists has not been rejected and continues to direct all sorts of quite expensive research, though the empirical evidence for such an hypothesis is meager and indirect. The prudential policy to adopt involves not merely the chance of being wrong but also how much inconvenience is caused by being wrong over a particular hypothesis. The full theoretical treatment of this situation is fairly complicated and is best developed in terms of the theory of games, the investigator and nature being regarded as playing against each other. Contrary to an earlier treatment by Wald, Braithwaite shows that nature's moves can be at random. It will be evident that, in this section of the book, however matter-of-fact it may be, we are not far from Pascal's Wager.

Practically all the first two-thirds of *Scientific Explanation* is devoted to the question of rejection of hypotheses. Most people would, however, feel that such a negative view is inadequate and that some grounds for acceptance as opposed to nonrejection must be given if a definite body of scientific theory is to be built up. This leads to a consideration of the inductive process. Braithwaite rejects those theories of induction which attempt to assimilate the process to deduction. In particular, he rejects certain statistical theories which start by assuming that a finite, if on occasion very small, prior probability can be assigned to any meaningful generalization that is not known to be false, and that this probability may be increased as each instance of confirmation is encountered. The rejection is on the grounds that there is no reason to suppose that an initial or prior probability of finite, if very small, magnitude can always in theory be assigned. The present writer is unconvinced, but can give no good reason for his lack of conviction.

Braithwaite's own belief in induction is essentially a pragmatic one, stated in a way derived from C. S. Pierce. A well-tested and unfalsified hypothesis is to be accepted because in the past the acceptance of well-tested unfalsified hypotheses has proved satisfactory in the development of science. Braithwaite is at great pains to demonstrate that this approach is not vitiated by a certain rather subtle kind of circularity.

The results are perhaps not entirely satisfactory, but this does not mean that the whole argument, which can only be rightly criticized by the few specialists of the same caliber as Braithwaite, is not extremely interesting and impressive. The book ends with a more or less neo-Humean analysis of causality, with a chapter justifying the use of teleological hypotheses in scientific systems and with a consideration of what we mean by explanation of scientific laws. The chapter on teleology should be of particular interest to the biologist; it is an eminently sensible corrective to the quite absurd though common position that practically implies all statements of purpose to be meaningless in a monistic universe.

VI. The Electronic Antichrist

Le danger n'est pas que l'âme doute s'il y a ou non du pain, mais qu'elle se persuade par un mensonage qu'elle n'a pas faim.

There is fundamentally only one way in which we learn anything, namely by experience and reflection thereon. The experience of each individual is private, but by making certain assumptions and adopting certain conventions a public account of parts of it can be given. Whenever the publicly communicable parts of experience can be organized by the use of theoretical constructions, we have what may be regarded as scientific learning. There is no essential difference between recognizing comparable topological relationships among the bones of the forelimb of a mammal and a bird, and recognizing formal similarities in the stories of the Holy Blood of Fécamp which floated onto the Normandy coast enclosed in Saint Joseph of Arimathea's gauntlet that had been sealed up miraculously in a hollow tree and of the finding of the body of Osiris by Isis after it had floated down the Nile in a coffer that later became miraculously incorporated in a tree trunk. In the first case the relationships have had their place in theoretical considerations of evolution and of the phylogeny of vertebrates, while a theoretical discussion on the veneration of relics would have to take cognizance of the relationships of the second case. In both instances there is a great deal more to be said; vastly more information is needed to reach a satisfactory theory. We are all reasonably certain that birds and mammals have a common ancestry; it is much less certain that the speculation that might be derived from the second case, namely that the veneration of relics is closely related to the Osiris cult of pre-Christian Egypt, would hold up on further investigation, though it would explain, for instance, the lack of reluctance to dismember the bodies of the saints, and perhaps the lavishness with which the dismembered parts were encased in gold.

From a formal point of view, it does not make any difference that one example is chosen from the field of comparative anatomy, the other from that of comparative religion. What one starts to investigate depends largely on his previous experience and present circumstances. It is indeed extremely difficult to think of any class of phenomena which would not yield some generalizations of interest if properly studied, but in every case the approach must be characterized by what medieval philosophers called *studiositas* as opposed to a merely trivial *curiositas*. Effort will always be needed, as it is very difficult to make a discovery. Most classes of phenomena which appear unimportant are likely to be in areas about which we cannot yet think of anything to ask. In every effort to learn, it is necessary to have imaginative insight, a reasonably critical and logical mind, a considerable range of relevant information, and a desire to explore the matter. If this desire to learn what can be seen of the truth is not present, it is unlikely that indirect rewards will achieve anything of importance.

About the mental equipment, and to some extent the knowledge, there is little argument. About the desire there is usually silence, or an explanation in terms of economics or psychoanalysis.

In dealing with biological processes, with individual human behavior, or with social interaction, it is customary to consider what are termed the mechanisms of such phenomena, by which is meant that a conceptual model or theory can be constructed, using a limited number of elements with well-established public meanings, which is shown deductively to produce the phenomenon under investigation. We have all seen during the past few years how physical models of mathematical technicians can be built which differ from the objects that they imitate in being constructed of metals, relatively stable dielectrics and gas, rather than of proteins and nucleic acids dispersed in a liquid phase, and in being very much larger and very much quicker. The properties of these models can give some idea of the way that the human brain may be operating during a calculation. It is also possible to make artificial models of living organisms behaving in a more generalized but more autonomous manner, feeding on the potential between the leads of some appropriate storage battery, which they seek when "hungry," and in other examples able to learn to run mazes like a laboratory rat. It is also known that a machine could be designed to seek from random situations particular examples that could initiate new behavior, and that owing to its great speed, this new behavior might be achieved well in advance of any nonmechanical human insight.

Since there also seems to be no reason why self-reproducing machines might not be made, the autonomous evolution of mechanisms becomes a distinct possibility, as has frequently been pointed out. As of the present time there is no danger of this happening, because the ordinary ecological principle of limiting factors would start to operate; various special metals, or even water for the cooling systems, very likely would not exist in adequate quantities. The limitation implied by the large size of existing computers, however, is not necessarily permanent. A great reduction in bulk has indeed recently been achieved by the use of transistors. Moreover, Ashby has suggested that if a single large machine were used to predict the effects of courses of action in political life, it would probably indicate, when faced with a long and complex problem, that the most efficient initial course would be to increase its size and complexity. The integrity and growth of the machine would in fact be the major concern of political life.

It is reasonably certain that most people would regard this kind of situation with great misgivings. It is also certain that a formal mechanistic analysis of such misgivings would indicate that there were prejudices, in the sense that if we regard the machine as a valid model in our analytical system, there is *within that system* no reason to prefer the welfare of a man to that of a machine. It is to be noted that a lot of other commonly accepted values would also appear in the same context as prejudices. Moreover, if the machine contained some mechanism tending to insure its integrity by indicating that it could give no useful answers if tampered with, and if it were farsighted enough to predict beyond what human foresight could conceive, it would start giving instructions about the use of language which well might finally suppress the use of any words giving information about its origin. The machine would then appear not merely indispensable but also supernatural. It is perfectly possible that in this way almost unwittingly a mechanical model of Satan might be constructed which would in fact satisfy most of the theological requirements of the Antichrist.

The above argument may appear fantastic, but it employs only reasonable theoretical postulates suggested by what is known of mechanical models and of evolutionary processes. The argument cannot be met by any glib reply that this sort of thing would never be permitted, because we now know that all sorts of horrors, which we might have supposed would never be permitted, have been permitted during the past thirty years in supposedly civilized countries. Rather it is cogent to ask under what circumstances would the electronic Antichrist come into existence if it were technically feasible. The writer suspects that two conditions would be necessary: firstly, an inability to distinguish between some given phenomenon and its mechanical model, so that the model comes to be a complete explanation of the former, rather than an explanation only of the properties the model was originally devised to explain; secondly, a preoccupation with the socially useful, so that a smooth-running, peaceful, wantless, egalitarian society is regarded as the main aspiration of mankind. Since it would be possible to make mechanical models of such a society, the two points of view resonate.

It seems likely that if the autonomous electronic Antichrist were imminent two quite different courses of action would be recommended to prevent its development. The first would be anti-intellectualism specifically directed against electronic machines, then by implication against mathematics and logic and so, by association, against the whole range of human mental activity. This, of course, would insure that whatever mechanisms, promising the millennium and producing tyranny, already might be available at a pre-electronic level, would come into full power that much earlier than their more fully developed electronic analogues.

The second course of action is to realize that, however successful our mechanistic explanations may be, we cannot know how much of reality we are incapable of knowing. This limitation is quite irrelevant when we are dealing with the scientific study of systems defined as external to the investigator; it is highly relevant when we are dealing with ourselves. Once we admit that there is an ineluctable mystery built right into us, which to the most skeptical must at least appear as the mystery as to whether there is such a mystery, we will be more careful about respecting desire, however well we can explain it.

The enormous prestige of applied science, and the overvaluation of the social aspects of human behavior, which can in theory be imitated to any degree of perfection by mechanical models, have led to a continued and growing neglect of the significance of science as a method of illumination which we know empirically that we can experience. Among philosophers, the only one with whose works the writer is familiar who has put the matter in a sufficiently uncompromising way is the difficult and sometimes seemingly perverse French saint, Simone Weil. She rejected completely the social significance of science as it is commonly understood. Her whole life was an intense and desperate exploration of a psychological wilderness of affliction, and from this wilderness she cried stridently that we should take in the entire universe in an act of intellectual love extending infinitely far into the future and past and excluding nothing but our momentary sins. It was as a part of this act that she conceived science, and for that reason this essay, ἐν ὑπομένῃ and in humility, is dedicated to her memory.[10]

Note in proof: A very beautiful account of computers as intellectual telescopes, by Margaret Masterman, has appeared in *Times Literary Suppl., 61* (1962) 284, and may be taken as a counterpoise to this essay.

10. The relevant passages are mainly in the last chapter of *L'Enracinement*, published in translation as *The Need for Roots*, 1952, and in the long essay (from which the quotation at the head of this contribution is taken) on the forms of the implicit love of God in *Attente de Dieu*, published in America as *Waiting for God*, 1951, both G. P. Putnam's Sons, New York. The title of the second of these books is, I understand, adapted from a phrase of the German existentialist philosopher, Martin Heidegger. In its American form it gives an entirely false impression of the contents of the work. The Greek words ἐν ὑπομένῃ (sic) mean *in patience* and were used by Simone Weil to express a state of waiting in absolute receptivity. In view of the fact that the various books and essays that have been published about her recently give hardly any indication of her attitude to science, and of the fact that her supposed affinities with existentialism have been specifically or implicitly stressed, it is necessary to point out that she was about as far from being anti-intellectual as it is possible to be, though some of her more speculative ventures into scholarship are doubtless original rather than sound.

NATURAL SELECTION IN THE GARDEN AND THE WHEATFIELD

Darwin, as is well known, considered that the multiplicity of races, exhibited by most animals and plants under domestication, provides a model of the multiplicity of natural species derived from a single ancestor. The existence of the model was in fact used as a convincing argument in favor of the evolutionary process in nature. Natural selection or the survival of the fittest took the place of the artificial selection of the most desirable. The eighteenth century had seen an extreme development of selection in the improvement of various domesticated species. Pantin,[1] in a fascinating address, has pointed out how large bets were made on the results. From the point of view of the plant or animal, the advent of the breeder was simply another factor added to the environment, a factor which sometimes made absurd demands, such as developing fantailed pigeons and the fantastic goldfish of China. Selection became very asymmetrical, but under domestication, as in nature, specialized environments could be regarded as producing specialized adaptations. Genetics and biochemistry might set limits; the late Lord Berners had to dye his pigeons pink and green rather than breed them, though outside the genus *Columba* plenty of tropical pigeons are green. On the whole, however, the plasticity of most species was remarkable and the model worked well.

Unfortunately, practically nothing was known about the origins

1. C. F. A. Pantin, "Darwin's Theory and the Causes of Its Acceptance," *School Science Review*, 3 (1950–51), 116: 75–83; 117: 197–205; 118: 313–21.

of most domestic species. All domestic pigeons, in spite of their extraordinary variety, were presumably derived from the rockdove, but when and where was unknown. Reasonable speculation on some other domestic animals was possible, but the origin of the important crop plants lay in complete obscurity. It is only in the last few decades that much progress has been made in such researches, and now it is the evolution of natural species that illuminates that of the domesticated forms. The processes involved, moreover, turn out to be most peculiar and very far from the conscious ennoblement that Darwin and his contemporaries believed to be the main factor in the process. Several works, namely Edgar Anderson's book, *Plants, Man and Life*,[2] I. H. Burkill's *Hooker Lecture, Habits of Man and the Origins of the Cultivated Plants of the Old World*,[3] and two papers of Helbaek[4] call attention to some of the more general problems in this field, and several special monographs have also appeared, one of which will be discussed below.

Man in his primitive food-gathering phase must have been an agent of natural selection producing results contrary to his own interests, as indeed he often does in his dealings with nature even today. Burkill points out that the Andaman Islanders, whose vegetable food is derived almost entirely from perennial plants, would have exterminated any yam that could be easily dug up. Their best wild tuber at the present time is *Dioscorea glabra* Roxb., and a religious ordinance is enforced preventing its being used during the season of new growth, when the local rain-god is said to need the tubers. Selection against ease of digging and perhaps against edibility must, however, continue.

By far the greater number of important food plants are annuals. Burkill emphasizes that the cultivation of perennials and above all of

2. Boston, Little Brown, 1952.

3. *Proc. Linn. Soc.*, *164* (1953), 12–42.

4. "The Paleoethnobotany of the Near East and Europe," *Oriental Inst., Univ. Chicago Stud. in Ancient Oriental Civilization*, *31* (1960), 99–118; see also "Domestication of Food Plants in the Old World," *Science*, *130* (1959), 365–72.

fruit trees is a relatively late invention often associated with hierarchical and urbanized societies.

Exactly where and when the beginning of true agriculture took place is as yet unknown. It is perhaps better at the present time to think of the beginnings of cultivation of particular plants or groups of plants. Anderson inclines rather to Carl Sauer's view that a fishing people who would have sedentary families in their littoral homes, however much they roamed the seas, built up accidental vegetable gardens on their refuse dumps, which would indirectly through excrement, if not directly through fish offal, be fertilized from the sea. Such refuse dumps would acquire what would be a typical weed community, pioneers on broken and fertile ground. Before man began disturbing the natural vegetation, such plants would be very rare. Some might be found on unstable flood plains. They would be typical fugitive species, easily dispersed and easily exterminated by their successors. A good many cultivated plants seem in fact to be derived from such weeds. We still eat several species of Chenopodiaceae, notably beet and spinach. Composites, such as wild lettuce, umbellifers such as Queen Anne's lace or wild carrot, and many legumes are now cosmopolitan weeds, while hemp or marijuana is notoriously easy to grow in abandoned city lots. Most of the dump and midden building that Sauer believes may have started agriculture began in the Mesolithic, but this period was not contemporaneous throughout the world. Much has been written about the localities where these activities first occurred. Vavilov, the great Russian student of cultivated plants, supposed certain specific and mainly montane centers in Ethiopia, in southwest Asia including the Caucasus, in India, in southern China, in the uplands of Mexico, and in Peru to be involved. The native cultivated plants of these regions show great variability, and might therefore have given rise to a vast number of later widespread descendants. Burkill thinks that this is to be attributed partly to the variety of microclimates in mountains and partly to the incompetence of mountain folk as agriculturalists. Vavilov's main thesis is probably erroneous. This, however, does little to detract from Vavilov's magnificent researches which will always be part of the foundation of the study of cultivated plants. Burkill himself thinks that a geographically peripheral species of a large genus is more likely to be improved under cultivation than one of a local complex of species which would be more liable to breakdown through hybridization. The known history of some species, such as the olive, is evidence in favor of this view, but in other cases, hybridization has been of major importance. Possibly Anderson's argument that native diseases flourish where many species of a genus are native explains Burkill's observations.

Anderson,[5] who has recently studied the neglected field of the origin of the horticulture of ornamental plants, thinks that there may have been a special development of the cultivation of plants for their decorative flowers in the Pacific basin, from which area the practice spread westward to Asia and Europe and eastward into Central America.

As soon as cultivation began, a new set of selection processes would be initiated. Burkill suggests that the earliest and latest maturing plants would be selected against, as the crop would be gathered when the majority of plants were ready. This would weed out any genetic combinations making for early or late maturation. At the same time, man would select against scattering mechanisms. Any effective husbandry, moreover, would minimize the need for various protective devices, awns that make grain hard for birds to eat, loose panicles that are hard for birds to perch in. Mutants that lacked these protective devices would be no longer selected against. If, as is possible, such mutants were more fertile or more easily harvested they would tend to displace the old types.

The most exciting recent work on the origin of any cultivated crops is certainly that of Helbaek, who seems to have caught both wheat and barley in the process of being domesticated, during the seventh millennium B.C. at Jarmo in the Kurdish hill country of

5. Edgar Anderson, "The Evolution of Domestication," in *Evolution after Darwin*, ed. Sol Tax, Vol. 2 (Chicago Univ. Press, 1960), 67–84.

Iraq. The material obtained in this region, during the studies of the Oriental Institute under R. J. Braidwood, consists of grain imprints on lumps of clay and of actual carbonized grains.

The cultivated wheats consist of a number of species. One, *Triticum monococcum*, is diploid ($2n = 14$) with a stiff glume over the grain that does not release the latter on threshing; this species is evidently derived from *T. aegilopoides*, known in the Balkans, Asia Minor, and Kurdistan. It is hard to hybridize with other wheats and, though *T. monococcum* was very likely mixed with other primitive wheat and barley early in their domestic history, it appears to have been of little importance. A large group of wheats is tetraploid ($2n = 28$), including the wild *T. dicoccoides*. The latter plant has a restricted distribution from Kurdistan westward toward the Mediterranean. It is evidently from *T. dicoccoides* that the first domestic wheat was obtained. Jarmo lies within the distribution range of the wild parent; the material found in the site is variable and intermediate between the wild species and the fully domesticated though still glumed, emmer or *T. dicoccum*, the wheat par excellence of the first farmers from Western Asia, Egypt, and northwest to Britain and Scandinavia. From *T. dicoccum* a number of threshable naked tetraploid wheats have evidently arisen. Helbaek feels that there is no good evidence for their existence before their use in Hellenistic Egypt. The most important of these is hard wheat (*T. durum*), which can be grown in steppe regions, dry in summer and unsuitable for bread wheat.

Most of the significant modern wheats are hexaploid ($2n = 42$). No wild hexaploid species is known, but spelt (*T. spelta*), one of the cultivated hexaploids, is glumed. Since it has proved possible to synthesize a form comparable to spelt by crossing tetraploid wheat with goose-face grass of the allied genus *Aegilops*, it is very tempting to suppose that spelt arose as an allopolyploid in this way, and from it the important modern hexaploid wheats, including the bread wheat *T. vulgare*, developed by mutation just as hard wheat is supposed to have done from emmer. Helbaek points out, however, that at present it appears that naked wheats having the grain shape of hexaploids (one cannot count chromosomes in a carbonized grain and still less in a clay impression) arose in Mesopotamia before 3000 B.C.; spelt, and incidentally the naked tetraploids, appear much later, the former seeming to develop in Europe, outside the range of the putative *Aegilops* parent. The problem of the history of wheat is thus still unresolved though good progress has been made.

The history of barley appears to be associated with that of wheat, as originally the two were grown indiscriminately together, though wheat predominates in the Jarmo material. As with wheat, the Jarmo barley appears intermediate between the wild *Hordeum spontaneum* of the Near East and cultivated two-rowed barley, *H. distichum*. In most features the Jarmo barley is like the wild plant, but there is clear evidence that the axis of the spike was tough so that, unlike the wild form, this early cultivated barley had ears that did not break to pieces when ripe, a mutation which would give great selective advantage to a plant that was reproduced by harvesting and planting rather than random dispersal. Again, as with wheat, the early history now seems better known that the later evolution of modern six-rowed barleys. As with other cultivated crops this later development presumably depended initially on natural selection operating in new ways on plant populations carried and tended by man under all sorts of ecological conditions quite different from those under which the wild parent species grew.

A very peculiar type of selection is believed to have given rise to modern oats (*Avena sativa*) and modern rye (*Secale cereale*). Oats seem to have started as a weed in emmer (*T. dicoccum*) fields in Neolithic or Bronze Age times. A very simple type of selection of seeds would save the larger, more easily harvested grain among the precursors of *A. sativa*, and finally the modern cereal was evolved as an adaptation to the sickle of the reaper, which insured a better chance of a new generation than did natural dispersal. Rye seems to have arisen in the same way in wheat and barley fields in Asia Minor during the Iron Age.

How far hybridization between species or even genera is a factor in the evolution of cereals may be uncertain, but the process is sufficiently well established in other plants for it always to be probable. The cultivated plant may, however, hybridize to give valueless crosses which masquerade as wild ancestors. *Euchlaena* is a well-known example in the case of maize; many tetraploid wild oats are known to be derived from the diploid cultivated form crossed with other grasses.

It is interesting in a general way to compare what is known of the domestication of animals with that of plants. Animal species are far less likely to produce autopolyploids or to hybridize, so giving rise to occasional allopolyploids with striking and often useful characters, than are plants. This is partly due to cytological causes, which may break down in hermaphrodite groups such as the earthworms, and partly to the fact that in early times only large animals interested man, and these for the most part belonged to genera with few or no sympatric species. Social behavior seems almost a prerequisite for domestication of animals. Of the older forms the cat, the guinea pig, and the silkmoth are perhaps the only exceptions. It is reasonable to suppose with Reed,[6] who has treated the whole subject with great learning in two recent summaries, that the process may have begun by women and children in hunting communities keeping as pets young animals whose mothers had been killed. The pet might be eaten when it became too large, or too unruly, or its owners hungry or bored. The first domestic animal was almost certainly the dog, followed in the Kurdish sites by the goat and pig at about the same time that wheat and barley were entering cultivation. At least in Western Eurasia, animal and plant husbandry grew up side by side in the seventh millennium B.C. If, as seems possible, the first wheat or barley fields were patches in oases that nomads planted deliber-

6. C. A. Reed, "A Review of the Archeological Evidence on Animal Domestication in the Prehistoric Near East," *Oriental Inst., Univ. Chicago Stud. in Ancient Oriental Civilization, 31* (1960), 119–45; see also "Animal Domestication in the Prehistoric Near East," *Science, 130* (1959), 1629–39.

ately, but left untended, in order to provide a ripe crop when they passed that way again, these nomads were on foot and not pastoralists, for the horse did not appear as a domestic animal anywhere until much later.

There is a general tendency for the early cultivated species to provide a number of products. Grain and straw both have their uses; hemp yields fiber, oil, and a drug; tumeric, used in curry powder, may at first have provided a ritual cosmetic. All the important plants have had ritual connotations. Wheat and the grapevine are built into the foundations of Christianity.

This story is inevitably abbreviated. It will be familiar to many readers, so familiar that they will find the above account entertaining only as a source of lacunae and perhaps errors. It is given to emphasize that, though initially artificial selection was a model of natural selection, now the situation is reversed. The origin of crop plants, and perhaps of some domestic animals, is a very special and complex case of evolution conceived in the broad sense. It is, however, also an aspect of the greatest human importance. Wheat, maize, tomatoes, yams, and potatoes are important factors in human history, and the migration of civilization has often followed a trail of perfumes and cosmetics. Anderson's book is in fact largely written to emphasize such things. It is a demonstration of what can be done by the taxonomic methods which he has invented, a challenge to use all such modern methods in studying a neglected subject and a plea for a far broader, more humane and more educated attitude toward plants and life. The book is of great educational importance, because its material can be used to bridge the dangerous gap that so easily can develop in high schools and colleges between the arts and sciences, or between the academic and the practical. It also may be of considerable importance in providing material in the education of aboriginal people in America, who have made great contributions to human welfare through their botanical discoveries and who should now be brought up to be conscious of, and proud of, this fact.

The argument that the great American cultures of ancient and medieval times were entirely indigenous developments has often been supported by the negative evidence that no important food crop was grown in Mexico and Peru that was also employed in Asia. It has been supposed that no people would carry with them memories of elephants, of terraced fields, musical scales, and board games who did not also have with them rice or some other stable agricultural plant. The argument is obviously not logically binding; memories are lighter cargo even than seeds and cannot be eaten in desperation en route. The absence of Asiatic plants in Central and South America, however, would be very curious if much traffic from Indonesia across the Pacific had really taken place. Statements regarding the absence of Asiatic plants have been made uncritically in the past. Detailed modern studies have indicated that there was some community of food plants between America and Asia. Moreover, it is very unlikely that such a community can be explained by transport through the island chain of the Bering Straits. The really curious thing about this modern evidence is that in the best cases it suggests movement from America to Asia, as Heyerdahl on the *Kon-tiki* would have it, and not from a center of high civilization in Asia into the American provinces, as has more often been postulated.

The cultivated cottons of the New World are allotetraploids, certainly derived by crossing an Old World diploid with a New World wild diploid species. The existence of such cottons may be, and often is, interpreted in favor of cultural diffusion from Asia. Since it now seems that the Old World parent is African in its distribution, most authorities think human migrations have not been involved and put the origin of the tetraploid back into pre-Pleistocene times when the Old World parent might have had a wider distribution. The case of the sweet potato, which spread from America through Polynesia and into New Zealand in pre-European times is, according to Anderson, quite definitely in favor of culture contact between the West Pacific and America; but this is not more surprising perhaps than the introduction of an Austral Island parrot into Fanning Island by visiting navigators in pre-European times. The most impressive and curious case is that of the grain amaranths, studied by Jonathan Sauer,[7] which provides apparently convincing yet puzzling evidence of diffusion of a grain crop into Asia from Central America in ancient times.

The grain amaranths belong to a section, *Amaranthotypus*, of the genus *Amaranthus*, that is certainly native to the New World, where there are a number of endemic species, some very restricted and others very widespread in distribution. No indication that this section has any native members in the Old World exists. Certain species of the section *Amaranthotypus* are cultivated for the sake of their brilliant red or orange inflorescences and are known to gardeners as prince's plume or love-lies-bleeding. The numerous female flowers of this inflorescence set small seeds, always black in truly wild forms, but often pale in the cultivated species. At least four species appear to have been cultivated for seeds in the New World, and at least two reached Asia.

The most important grain amaranth is *A. leucocarpus* Wats., which is still grown sporadically in many parts of Mexico. Nowadays it is used mainly to produce a confection known as *alegria*, which is composed of the seeds, popped by toasting, mixed with honey or caramelized syrup. The ground seeds are also used mixed with milk or water as a drink. Everywhere the crop has the appearance of being an unimportant survival, which, if the women did not remind the men to plant it, would die out. In some places only a part of the population realizes its existence. Nevertheless, its area of dispersal is very great. Scattered cultivation is known in Guatemala, where the species used is *A. cruentus* Linn., a species also sporadically cultivated in Mexico. In Peru and Bolivia *Amaranthus caudatus* is mainly grown, while in the western Argentine, *A. edulis* Spegazzini is not only known today but has been found associated with other food crops in an archaeological site.

7. "The Grain Amaranths: A Survey of their History and Classification," *Ann. Missouri Bot. Garden*, *37* (1950), 561–632.

Sauer has produced a great deal of evidence, all fragmentary, but having great cumulative force, that a mysterious grain called *huanhtli*, or *bledos* by the Spaniards, which is usually taken to be *Chenopodium nuttalliae* or *C. pueblense* and so mistranslated as *beet*, was actually *A. leucocarpus*. This is a matter of importance, because huanhtli was of great ritual significance. It was a major ingredient of the dough, mixed with honey, of which an image of Huitzilopochtli, the Aztec war god, was made at his feast. This image was broken up at the end of the day and the pieces were distributed to the people to be eaten sacramentally as the flesh and bones of the god. Numerous similar rituals of less importance evidently existed.

Sauer interprets these facts as suggesting that *A. leucocarpus* was one of the primitive foodstuffs of Mexico. In immediate, post-Columbian times it had become less important in most places than maize, and is now lingering only as a survival among people with a conservative taste in sweetmeats.

In Asia the main center of cultivation of grain amaranths is in the mountains of northern India, though culture is also known in Ceylon. Virtually nothing is known about the history of the plant in these countries. Further west in Szechwan there is another, though possibly not disjunct, center of cultivation. In this area the plant is termed *tien-shu-tze* or *jen-shien*. Though historic data on these plants in the Orient are almost absent, a species termed *jen-hien*, meaning amaranth as high as a man, is mentioned in a tenth century pharmacopeia written for a prince of Szechwan. In its Chinese home, amaranth grain is popped and made into candy, much as in Mexico. The really surprising thing is that these Asiatic amaranths, which have all the appearance of being of ancient origin in the Orient, prove to be taxonomically indistinguishable from *A. leucocarpus* and *A. cruentus*. The range of variation is smaller in Asia than in America but every Asiatic variant can apparently be matched by one in the New World. There can, in fact, be no doubt that the grain amaranths of Asia are American, but how and when they got to India and China is a mystery. In the past, archaeologists have been notoriously careless in preserving what seemed to be valueless. Renewed investigation may ultimately provide material that has been entirely overlooked and so settle a historic problem that is obviously of universal interest.

A SPECULATIVE CONSIDERATION OF CERTAIN POSSIBLE FORMS OF SEXUAL SELECTION IN MAN

The present contribution,[1] necessarily based on very incomplete information, is presented mainly because the subject appears to be important and because the implications of such data as exist seem to be little realized even by investigators in the fields to be discussed. It is hoped that its publication may stimulate research.

The total population of one generation of any organism is not derived equally from each member of the preceding generation. A certain number of individuals in any generation will be less fertile, others more fertile, than the mean. If any genotypes are more frequent among the less fertile portion of the population than in the more fertile portion, it is evident that one of three things must be happening: (1) there may be selection against the genotypes in question so that they become rarer, or (2) they may be maintained in the population by mutation in the face of selection, or (3) selection in favor of heterozygotes, or some other form of balanced polymorphism, may occur in the population.

1. A preliminary statement of the fundamental idea was made in a somewhat generalized form in an earlier review of contemporary work on human evolution (*American Scientist*, 45, 1957, 86–96). At the time of writing this earlier paper, I did not know of F. J. Kallmann's remarkable work on homosexuality in twins, which has led me to reconsider the entire problem. I am indebted to David Lack, D. F. Poulson, and E. S. Deevey, who read earlier drafts of the paper, to my wife for much helpful criticism, and particularly to Lawrence S. Kubie, who subjected all my ideas to a most generous and intense critical evaluation. Almost no one agrees with the entire paper, but no critic believes it without value. The initial exploration of the problem was made during tenure of a John Guggenheim Memorial Fellowship.

The human populations of Europe and North America, and probably of other parts of the world, contain a small group of people overtly exhibiting what is best termed *paraphilia*, namely a tendency in their sexual behavior to substitute, in the place of a partner of the opposite sex, goals which cannot lead to reproduction. The two most important forms of paraphilia are homosexuality and fetishism. According to Kinsey, Pomeroy, and Martin,[2] approximately ten per cent of the males of reproductive age in the United States exhibited only (group 6) or mainly (group 5) homosexual behavior for at least three years, at a time prior to interview. Among women, the incidence is supposed to be much lower, amounting to a proportion of the order of one per cent.[3] With regard to fetishism, the same authors merely say that individuals "who respond only or primarily to objects which are remote from the sexual partner, or remote from overt sexual activities with a partner, are not rare in the population." The phenomenon in its patent form is usually regarded as almost exclusively masculine. It is evident that, although the data are imperfect, if any genetic basis for these conditions exists, they are common enough to play a considerable part in the demogenetics of the population as a whole.

At present, any conclusion as to lowered fertility in paraphilia is largely a presumption, though a reasonable one. The data of Kinsey and his co-workers obviously cannot allow for possible changes in behavior after their interviews, so that to assume that all the group-6 homosexuals (exhibiting only homosexual behavior) are nonfecund is erroneous. Kallmann[4] found that in a population of

2. A. C. Kinsey, W. B. Pomeroy, and C. E. Martin, *Sexual Behavior in the Human Male* (Philadelphia and London, W. B. Saunders, 1948).
3. A. C. Kinsey, W. B. Pomeroy, C. E. Martin, and P. H. Gebhard, *Sexual Behavior in the Human Female* (Philadelphia and London, W. B. Saunders, 1953).
4. F. J. Kallmann, "Comparative Twin Study of the Genetic Aspect of Male Homosexuality," *J. Nerv. Ment. Dis.*, *115* (1952), 283–98; "Twin and Sibship Studies on Overt Male Homosexuality," *Amer. J. Human Genetics*, *4* (1952), 136–46.

85 twin homosexual men of median age over 26, all rated as belonging to group 3 or higher (that is, at least as homosexual as heterosexual, or more so), and only 11 had contracted marriages. These marriages were mostly short-lasting. Three of the marriages had produced children. In no case of the five children born is the biological paternity certain beyond a reasonable doubt. This fertility looks very low, but cannot be properly evaluated without more data on age structure. Davis[5] observed a supposedly significantly greater proportion (38.8 per cent) of childless marriages among married women who admitted to postmarital homosexual activity than in her over-all sample (20.8 per cent). The average age at the time of reply was 38.2 years in the case of the small (18) homosexual sample, 38.3 years in the case of the large (1,000) over-all sample. Such meager evidence supports the common-sense view that homosexuality in Western cultures is likely to be correlated with a low fertility.

With regard to fetishism, there is no information of particular value. The extreme cases reported in the clinical literature are often obviously psychologically sterile, but there are also doubtless plenty of men with markedly fetishistic tendencies who achieve normal-sized families. One may presumably assume continuity between the two groups, which of itself implies a somewhat lower fecundity, though it is not possible even to guess how great the effect might be.[6]

Though it seems evident that various forms of paraphilia are common enough to be significant in demogenetics and that part of this significance is likely to depend on lowered fertility, any importance that the condition may have in the genetics and evolution of human populations will be dependent not only on frequency and

5. K. B. Davis, *Factors in the Sex Life of Twenty-two Hundred Women* (New York and London, Harper, 1929).

6. Kubie (*in litt.*) suggests that in the cases of "bisexual" homosexuals, or fetishists who marry, there may be an unconscious tendency to avoid coitus during the more fecund parts of the menstrual cycle, so reducing fertility.

low fertility but also on some degree of genetic control. At the present time, this is the aspect of the problem that is hardest to treat.

Some introductory remarks on the general problem of the inheritance of temperamental differences are necessary before proceeding to a discussion of the one body of work that is relevant to the details of the present discussion. The pioneer students of sexual as well as other kinds of psychopathology in the late nineteenth century usually assumed an hereditary predisposition to account for any condition whose genesis they did not understand. The high selection rate that would obviously be operating against some of the supposedly inherited conditions that they considered seems not to have worried a generation that knew nothing of the magnitude of mutation rates nor of the mathematical theory of selection in natural populations.

During the present century, the ascendency of psychoanalysis in clinical psychiatry and of various types of behaviorism and learning theory in academic psychology have led to a tendency to dismiss all genetic causes of variation in human behavior. This has probably happened for several reasons. Firstly and most respectably, it might be claimed that since fairly satisfactory explanations of a great variety of different types of behavior have been given in terms of upbringing and experience, no hypothesis of genetic control should be introduced unless it can be shown to be necessary. So far as psychoanalytic procedure goes, it is probably safe to say that the ordinary techniques employed are *per se* not logically capable of disclosing any facts that necessitate a genetic explanation; individual analysts, for reasons outside these techniques, may have suspected constitutional factors[7] to be operating in particular cases, though these factors are not necessarily genetic. Secondly, the academic psychologist is well aware of what are commonly called species differences, as for example between the pigeon and the white rat, to name only examples from different classes. A sufficiently large array of similarities in diverse organisms has perhaps led to a tendency to under-

7. By constitutional in such cases is meant genetic factors together with such environmental factors as have operated prenatally.

rate specific and individual differences, which in any case are not wanted in statistical experiments. Thirdly and less respectably, environmental determination has seemed more amenable to manipulation and, therefore, more optimistic (from the standpoint of the manipulator). Lastly, in the past two or three decades, it is probable that feelings of guilt have become associated with hatred of particular scapegoat groups; since most psychologists are decent and humane men, the hatred has been repressed and as a reaction all theories of genetic labeling of distinguishable groups of men have been suspect.

When we examine the evidence that exists, the results are very difficult to evaluate. The most critical procedure consists in the statistical comparison of identical (uniovular or monozygotic) twins reared from infancy in different environments, as compared ideally with fraternal (binovular or dizygotic) twins also separated at infancy. The pioneer study of Newman, Freeman, and Holzinger[8] suggested, from a study of separated uniovular twins, that morphological characters were least likely to be altered by ordinary environmental influences, that formal intellectual characters as measured by intelligence tests were somewhat more modifiable, and that certain aspects of personality were quite easily modified. Neither their data nor those of other studies exclude genetic control of variation of personality, though they do seem to indicate that such control is of much less importance than in physical or formal intellectual characteristics. Criticism of this work in terms of the results of studies of twins reared together, as has sometimes been made, is quite improper and misleading. The intellectual justification of an extreme position in favor of almost complete environmental control of personality may therefore appear correct if we examine only human psychological data. Viewed against a wider background, it seems much less satisfactory.

Throughout the animal kingdom there are examples of genetic

8. H. H. Newman, F. N. Freeman, and K. Holzinger, *Twins: A Study of Heredity and Environment* (Chicago Univ. Press, 1937).

control of every kind of function including behavior, while in man there are good examples of genetic variability affecting every sort of character for which clear-cut evidence can be obtained. We are, therefore, practically obliged to assume that, if traits of personality in the widest sense are not in part genetically determined, there is some mechanism operating to screen the learning human being from the effects of genetic diversity. The existence of such a screen would be all the more remarkable in view of the inadequate screening provided against learning the wrong thing, or the right thing at the wrong time. Data such as those of Newman, Freeman, and Holzinger do suggest that such a screen is in fact partly developed; even in an imperfect form, its existence is obviously a very important datum for the student of human evolution. At least the present writer would find the assumption of an almost perfectly developed homeostatic mechanism[9] of this kind a more difficult assumption than to believe that there is some genetic determination of variation in affective behavior, even if psychotherapeutic techniques cannot by their very nature tell us of the existence of variation in the genetic background. Some degree of behavioral independence from genetic control may perhaps be implied by the general adaptability of mammals as compared with other vertebrates. A very large brain could not develop accurately enough to contain much built-in detailed information.

We may now turn as unprejudicially as possible to the main study of modern times suggesting a strong genetic determination of paraphilia. Kallmann has obtained data for the twin brothers of 40

9. It is to be noted that the existence of homeostatic mechanisms always gives the initial impression of false simplicity. Extremely complicated behavior, such as that of man, almost certainly implies learning, because to make a nervous system capable of really elaborate instructive acts of many kinds would probably put too great a strain on the conservation of information as the system was developed by cell division. The major structural bases of behavior are, however, brain structures, determined genetically, and exhibiting a degree of variation comparable to that in other organs as far as is known.

uniovular and 45 binovular homosexuals all rated as in groups 3–6. In the binovular group, there were 26 male twin sibs, of whom 15 were entirely heterosexual, 10 exhibited moderate homosexuality (group 1–4), and only one was almost exclusively homosexual (group 5). These proportions are not significantly higher than in the population at large. Of the uniovular sibs, three were deceased or unavailable, all the other 37 were homosexuals of groups 3–6. Nineteen uniovular homosexuals of group 6 had twin brothers rated in groups 5 and 6; the twin of the twentieth member of this group was unavailable. Of the ten uniovular members of groups 3 and 4, nine twin brothers were available and were rated as belonging to groups 3 to 5.

Since these data refer for the most part to individuals reared together in infancy and to a varying unspecified age, they do not have the critical significance of studies on twins reared wholly apart after adoption.

The main difficulty in accepting the work lies in the possibility that being raised as a twin might in itself predispose an individual to homosexuality. Dr. Kubie (*in litt.*) points out that a prohibition not to play with one's own genitalia might be interpreted by a twin child as permission to play with the genitalia of the other twin. Mutual homosexual seduction might, therefore, be particularly likely in unisexual pairs of twins. Such a process would ordinarily have been forgotten at the time of interview in adult life and is not, therefore, inconsistent with Kallmann's emphatic statement that none of his twin pairs had ever engaged in homosexual practices *inter se*. It is possible that the somatic identity of the two members of a uniovular pair might facilitate such a process, but it is by no means obvious that it would produce the very great difference between the identical and fraternal pairs recorded by Kallmann. The latter does not give any details of the criteria used in determining whether his twins were mono- or dizygotic, nor does he say whether the criteria were applied before or after the rating for homosexuality had been made. There may have been some opportunity for the operation of unconscious subjective factors. The writer is inclined to accept the findings provisionally as indicating at least some genetic determination, though probably not as much as would appear at first sight. An investigation of the incidence of homosexuality among single uniovular and binovular twins would probably indicate whether twinning *per se*, and in particular uniovular twinning, promoted the condition. This would doubtless be easier than the extremely difficult procedure of Kallmann. Such an investigation would throw no light on the problem independent of Kallmann's data, but would tend to validate the latter if no differences were found. The still easier study of marriage rates, if the samples were large enough to compensate for inevitable loss of information, might provide some significant results. As far as the writer can determine, even this has not been done.

For fetishism we have no comparable data. Greenacre[10] says she is not impressed by the possibility of any constitutional disposition toward the condition, but the analogous point of view would have been taken by many psychoanalysts discussing homosexuality prior to or even in view of Kallmann's work.

The only way of reconciling Kallmann's findings with the inevitable high rate of selection against the genotypes apparently partly responsible for paraphilia is to suppose that the modal type in the population is genetically unstable and that the individuals of presumed low fertility predisposed to paraphilia represent largely homozygous recombinations which are constantly being eliminated without changing the average genetic constitution of the population. To suppose that a process of progressive elimination of paraphilia is occurring would hardly be in accord with historic evidence, and would raise the difficulty of providing an explanation as to how the genotypes involved reached their former high frequency. The proportion of the population concerned seems to make a hypothesis of mutation exceedingly improbable.

10. P. Greenacre, "Certain Relationships between Fetishism and Faulty Development of the Body Image," *Psychoanal. Study Child.*, *10* (1955), 189–94.

The obvious model implies a situation not unlike that which Penrose[11] believes to underlie the population genetics of intelligence. The model is a special case of that developed by Goodhart considering all possible types of genetic infertility. It bears formal similarities to the much more extreme type of determination of the fertile caste in the bee *Melipona* in which queens in most species must be heterozygous for three genes.

The classical psychoanalytic theory of the development of homosexuality in males involves an attempt on the part of the growing child to solve the conflict of the Oedipus situation by identifying with the mother and so in effect replacing the rivalry of the father by love directed by the father to the mother. One gets the impression from the literature that this is most likely to happen when there is strong maternal dependence in the pregenital and early genital stages of development.

The classical psychoanalytic theory of fetishism is that it represents an attempt to deal with the male child's fear of castration, which he develops when he becomes aware of the anatomical difference between the sexes. The fetishist is thus a man who unconsciously tends to reject his early anatomical discovery, denying that women really lack a conspicuous external intromittent organ. The fetish is in fact a substitute for the missing female phallus.

Recent studies, admirably reviewed by Greenacre to whom much of our modern knowledge of the subject is due, suggest that extreme fetishism not only involves rather severe traumatic experience in the early genital stage of development (two to four years) but that

11. L. S. Penrose, "Evidence of Heterosis in Man," *Proc. Royal Soc. London (B)*, *144* (1955), 203–13. Compare also C. B. Goodhart, "Natural Regulation of Number in Human Populations," *Eugenics Review*, *47* (1955), 173–78, and "The Inheritance of Intelligence," *New Scientist*, Aug. 22, 1957, and "The Future Human Fertility," *New Scientist*, Dec. 12, 1957; G. E. Hutchinson, "Marginalia (Homo vosce te ipsum)," *American Scientist*, *45* (1957), 88–96; W. E. Kerr, "Genetic Determination of Castes in the Genus *Melipona*," *Genetics*, *35* (1950), 143–52, and "Evolution of the Mechanism of Caste Determination in the Genus *Melipona*," *Evolution*, *4* (1950), 1–6.

traumatic experience, usually fever, nutritional disturbances, and the like between the ages of 0.5 and 1.5 years, have made it difficult for the developing infant to establish an adequate image of its own body. There is concomitantly a great deal of clinging maternal dependence, and identification with the mother or other female member of the household. The highly disturbed cases who come for analysis frequently exhibit homosexual and sadomasochistic tendencies. Greenacre concludes that the various forms of paraphilia are closely interrelated in origin. Both homosexuality and fetishism involve a partial or complete rejection of a realistic view of women; in both cases there seems to be an early excessive dependence on and identification with a woman. Such a conclusion is probably valid even if some of the details as to how the conditions arise are still controversial.

It is important to realize that latent homosexual and fetishistic tendencies always seem to be present in human beings. Any genetic mechanism must be involved only in determining whether they are of such a kind as to lead later to particular kinds of action.

The best way of reconciling Kallmann's findings with those of the psychoanalysts (whose theoretical formulation Kallmann himself, for reasons that seem to me inadequate, appears to distrust) is to suppose that the genotypes responsible for a tendency to paraphilia operate primarily on the rates and extent of development of the neuropsychological mechanisms underlying the identification processes and other aspects of object relationships in infancy. There is more than a suspicion that temporal lobe function would be involved in such a hypothetical mechanism.[12] Such a hypothesis would be in accord with a great deal of modern embryological genetics, in which the genetic control of rates of various processes rather than the

12. A case of a man who was sexually excited by safety pins until after a small temporal lobe operation for focal epilepsy, after which his erotic life became normal, is reported by A. W. Epstein in "Fetishism: A Study of Its Psychopathology with Particular Reference to a Proposed Disorder in Brain Mechanisms as an Etiological Factor," *J. Nerv. Ment. Dis.*, 130 (1960), 107–19.

control of the mere existence or nonexistence of the processes is usual. Along these lines, it is to be noted that the variations to be explained are likewise not the clear-cut presence or absence of a trait, but its greater or lesser expression in action, when it may always be present as a verbally reportable subjective fantasy. The hypothesis also has the great merit of leaving the psychoanalytic theory, and in favorable cases the possibility of psychotherapy, untouched, though it provides a frame within which the theoretical model operates.[13] The probability of a number of different genotypes being involved fits well with Kallmann's impression of the considerable diversity of his material. It may also be noted that this diversity need not necessarily be expressed in homosexual or other paraphilic behavior. Some of the genotypes may exhibit no phenotypic differences, and there is a vast range of variation in such behavior, not necessarily involving lowered fertility, that might be in part determined by the same system.

The suggestion that a rather basic type of psychological development is involved, which is likely to have pleiotropic effects in nonsexual behavior, is in line with Kallmann's belief, if his somewhat arcane expressions are correctly understood by the writer, that the observed phenomena represent dislocations of a relatively lasting nature of the whole personality structure. In his particular cases, only 18 out of 85 were diagnosed as sufficiently adjusted both socially and emotionally, but it must be remembered that twinning and the

13. The at one time very popular theories of endocrine disturbance, and the more extreme view that homosexuals are cytologically and psychologically female but somatically male, now seem very dubious. The former has been considered and found wanting by Ford and Beach (*Patterns of Sexual Behavior*, New York, Harper, 1951); the latter, which rests on equivocal findings relating to the sex ratio, has to be supplemented by *ad hoc* hypotheses explaining why other inherently less improbable types of intersexuality are so rarely observed. This hypothesis appears to be falsified by the fact that individuals showing the Kleinfelder syndrome are in fact somatically male but cytologically XXY and appear ordinarily to be normal males in behavior.

social pressures against homosexuality may well produce lack of adjustment independently of the genotype.

From a more general point of view, the main importance of this type of selection, if it exists, is that it is operating on developmental processes which must be almost exclusively human and have only rather remote analogues among the lower animals; the developmental processes, moreover, are such that, as has just been indicated, they are likely to have effects on the personality outside the specifically reproductive and sexual fields.

The main objection that can be raised against the present hypothesis is that in some societies homosexuality is said to be unknown, while in others it is said to be a universal form of extrareproductive male activity. Actually, what we are really concerned with is the incidence of genetically determined psychological failure of functional heterosexuality. The permitted or institutionalized homosexuality of some cultures may involve far less rejection of the female than is found in homosexuals in Western Europe and North America. It may rather be compared to the apparently regular homosexuality exhibited in nature by some mammals, notably giraffes.[14]

The problem, from a cross-cultural point of view, is really whether for example there exist among the Siriono, who are said almost never to practice homosexuality, or among the Aranda, where nearly all men are said to engage in homosexual practices, a small group of men psychologically unable to achieve fertile heterosexual unions. In our own society, extreme homosexuality is apparently a frequent concomitant of this; in other societies, the same kind of selection might take place in a somewhat different way. I have failed to find any information as to whether in societies in which homosexuality is supposedly absent there are anerotic men comparable to the anerotic (group X of Kinsey's studies) women in ours, or whether in societies tolerant of homosexuality some men are never heterosexual.

14. A. C. Innes, "The Behavior of the Giraffe, *Giraffa camelopardalis*, in the Eastern Transvaal," *Proc. Zool. Soc. London*, *131* (1958), 245–78.

It has been suggested that the various idiosyncratic preferences that influence men in falling in love represent the normal expression of a phenomenon of which the pathological form is extreme fetishism.[15] This point of view evidently either rejects the psychoanalytic theories of fetishism, or takes no account of the great importance of the human face as the first visual stimulus to which a definite response is given by the developing infant.[16] There are, however, a certain number of specific and usually visual stimuli supplied by artifacts which tend to have erotic significance, which can become the objects of extreme fetishism in pathological cases, but which are so widely used in some societies that they evidently play a normal if minor role in human behavior. In contemporary society, high-heeled shoes provide an obvious example; in the fifteenth century, the tall conical headdress with a veil, known as the hennin, was almost certainly another.[17]

The significance of this sort of ornament is presumably to neutralize almost universal homosexual tendencies and castration fears which might otherwise impede reproduction. As such, it appears erotic and feminine, in spite of its hidden masculine significance. The term *cryptandric* is suggested to designate such ornaments and the behavior involved in their use. Cryptandric behavior may involve conscious modification of somatic characters, as in the foot-binding formerly practiced on upper-class Chinese girls. It is possible that a number of ornaments and practices employed by primitive as well as literate peoples are to be regarded as cryptandric;

15. A. Binet, "Le Fétichisme dans l'amour," *Rev. Phil.*, *24* (1887), 143–67. V. W. Grant, "A Fetishistic Theory of Amorous Fixation," *J. Soc. Psychol.*, *30* (1949), 17–37.
16. R. A. Spitz, and K. M. Wolf, "The Smiling Response: A Contribution to the Ontogeny of Social Relations," *Genet. Psychol. Monogr.*, *34* (1946), 57–125.
17. It is said that 20,000 women burnt their hennins as the result of the preaching of the French Carmelite Thoman Conecte in the first third of the century. After he was executed for heresy involving denunciation of clerical celibacy, the hennin returned to fashion for several decades.

this is particularly likely of ornaments worn through artificial holes in the body. Darwin[18] quotes from Livingstone a conversation between the latter and Chinsurdi, chief of the Makololo in South Central Africa. When Livingstone asked why women wore a ring (*pelelé*) in the pierced upper lip, Chinsurdi replied, "For beauty, they are the only beautiful things women have; men have beards, women have none. What kind of a person would she be without the pelelé? She would not be a woman at all with a mouth like a man but no beard." There is also an evident connection with certain kinds of charms used as protection against the evil eye, which are often overtly sexual though either male or female.[19] Psychoanalytically, the fear of castration is supposed to underlie the fear of the evil eye. It is interesting that Greenacre considers that an excessive dependence on charms and other lucky objects is a feminine counterpart of masculine erotic fetishism in contemporary society.

Anyone who has recently examined his copy of Darwin's *Descent of Man* will have been struck by the curious disparity between his account of structures believed by him to be produced by sexual selection in nonhuman animals and his account of supposedly like structures in man. The animal cases mainly refer to epigamic characters in the male, which are now understood to be either specific or sexual recognition marks or to be used in threat-displays, often of a territorial kind. The human cases are very diversified, largely female, frequently artificial and of no obvious functional significance.

It is reasonably certain that sexual selection in man, if it has been the sort of evolutionary force that Darwin believed, has largely involved the face. It might also be reasonable to suppose that certain exaggerated female structural modifications, such as steatopygia, function in a cryptandric manner. A somewhat intermediate case between artificial cryptandric modification and such a genetically

18. C. R. Darwin, *The Descent of Man and Selection in Relation to Sex* (1st ed., 2 vols., 1871; 2nd ed. 1 vol., 1874; London, J. Murray).

19. See above, pp. 4–9.

determined structure is provided by the umbilical hernia, not uncommon among the Bantu people in parts of southern Central and northern South Africa, which the writer was told in 1926 was regarded as very attractive in a girl. The resulting rather phallic protrusion obviously bears some resemblance to the scarifications so commonly used in Africa. Doubtless such a structure is one that could often be avoided by appropriate care during parturition, if it were not considered desirable.

If a certain amount of female cryptandric display, either structural or artificial, is very widespread in man, as seems possible, it may serve to neutralize the psychoanalytic effects of developmental rates, optimal in other ways, but reproductively suboptimal in the absence of the display in question. This may explain the apparent differences in the epigamic character of man and most other animals.

THE DODO AND THE SOLITAIRE

> The progress of Man in civilization, no less than his numerical increase, continually extends the geographical domain of Art by trenching on the territories of Nature, and hence the Zoologist or Botanist of future ages will have a much narrower field for his researches than that which we enjoy at present. It is, therefore, the duty of the naturalist to preserve to the stores of Science the knowledge of these extinct or expiring organisms, when he is unable to preserve their lives; so that our acquaintance with the marvels of Animal and Vegetable existence may suffer no detriment by the losses which the organic creation seems destined to sustain.

Thus wrote Strickland in the introduction to the classical memoir[1] on the dodo, published in 1848. It is both extraordinary and tragic how little we still know about animals which have become far rarer than they were at the time Strickland wrote. No satisfactory treatment of *L'Éléphant et ses amours*, let alone a stately ten-volume

1. H. E. Strickland, and A. G. Melville. *The Dodo and Its Kindred, or the History, Affinities and Osteology of the Dodo, Solitaire and Other Extinct Birds of the Islands Mauritius, Rodriguez and Bourbon* (London, Reeve, Benham and Reeve, 1848, 141 pp.). The copy examined in the Yale University Library contains an inscription, "To Professor Silliman with the kind regards of the authors," and a pencil note below: "Rec'd Octob 23–1848. The first part read and the second looked through within 8 days—." Beneath this is the signature in ink of James W. Dana, doubtless added later.

Grundriss der Elefantenlehre, appears, in spite of the popular story, yet to exist. We know even less about the five existing species of rhinoceroses, though now only one can boast a population of more than a few thousand individuals. The little that we do know is, however, abundantly worth knowing. Of the dodo we cannot now learn much that is new; what is ascertainable will be discussed in the following pages.

No animals that have ever lived seem to have balanced more precariously on the boundaries of the real and the imaginary than the flightless birds which are placed scientifically in the family *Raphidae*. Now, unfortunately, no one can observe their behavior. They lived on the islands of the Mascarene group, Réunion, Mauritius, and Rodriguez. The dodo of Mauritius, as everybody knows, has also been an inhabitant of Wonderland since 1865, the very year in which the best skeletal remains of the bird were discovered and sent to Europe. The solitaire never successfully left its home in Rodriguez, though there is a suspicion that before its demise it adopted some of the manners of the court of Louis XIV. The birds from Réunion, without bones to represent them, are solely known from the reports of a few travelers and from certain problematic pictures and may, in part, be fabulous.

The existing information about these strange creatures has been summarized by the late Viscount Hachisuka, who died a few days before his work appeared.[2] His book provides an admirable excuse for an excursion into one of the oddest regions of natural history.

The three islands which were the homes of the birds under consideration lie east of southern Madagascar, as shown in Plate 2. They were probably known to the Arabs but were first sighted by

2. Masauji Hachisuka, *The Dodo and Kindred Birds or The Extinct Birds of the Mascarene Islands* (London, H. F. and G. Witherby, 1953, 250 pp.). The book is primarily a compilation, and at times somewhat naïve; its value to the nonspecialist may, however, be judged from the fact that although one of the largest libraries in the New World is available to the writer, he had to rely on Hachisuka for the content of the works of Oudemans, Mortensen, Vivielle and Dehérain, and Killerman in preparing this account.

White Dodo of Reunion

Dodo of Mauritius

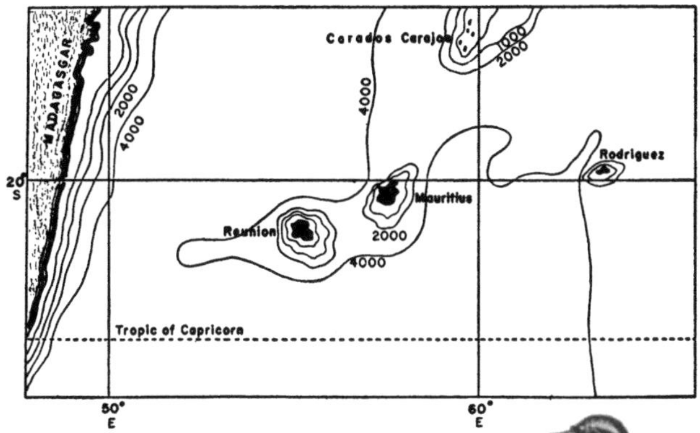

White Solitaire or Gaunt White Dodo?

Solitaire of Rodriguez

PLATE 2. Dodos and solitaires, with a map of the Mascarene Islands. (Drawn by Martha M. Dimock from various sources.)

European navigators early in the sixteenth century. All were well forested, lacked land mammals other than bats, and supported a number of birds, mostly now extinct, as well as giant tortoises. Of the last named, three species are supposed to have lived on Rodriguez and a considerable number, perhaps not all contemporary, have been described from subfossil or fossil material from Mauritius. Hogs, goats, and chickens were liberated on Mauritius in the sixteenth century, but the island remained uninhabited until 1638 when the Dutch established a settlement. Réunion was apparently visited for timber early in the seventeenth century and was inhabited sporadically after 1649. Rodriguez was supposedly first colonized in 1691 by a small group of Huguenots who attempted to establish a settlement.

Though no skin of the dodo still exists—Oxford University having destroyed the only specimen as too moth-eaten in 1755—the Mauritius species, *Raphus cucullatus* Linnaeus, is very well known. Over twenty-five illustrations made during the seventeenth century are in existence; these seem mostly to have been executed from captive birds brought to Europe or in one case to India.[3] Some of these

3. The charming Indian painting recently discovered in Leningrad has been reproduced in color by Erwin Stresemann, *in* "Wie hat die Dronte (*Raphus cucullatus* L.) ausgesehen?" *J.f. Ornithol.*, *99* (1958), 441–59, with a learned discussion of the circumstances of its execution and its significance in gaining an idea of the appearance of the bird. Herbert Friedemann, *in* "New Light on the Dodo and Its Illustrators," *Smithsonian Report, 1955* (1956), 475–81, has reproduced, among other drawings, one of three dodos now in the Crocker Art Gallery, Sacramento, Calif., which is apparently an original study of Roelant Savery. The feet of the birds depicted, as those of his finished paintings, appear slightly webbed, which from the surviving foot at Oxford and Peter Mundy's explicit testimony is known not to have been the case. Friedemann therefore suspects that Savery's opportunities to study the dodo were less good than has been supposed and that the accuracy of his finished paintings is somewhat questionable. He also shows that a drawing used to illustrate Bontekoe's account of his voyage, in which the white dodo of Réunion is mentioned, is only a poor reversed copy of one of Savery's drawings and cannot give any evidence about the Réunion bird. It is evident that too much reliance must not be put on the minor details of the contemporary drawings and paintings.

pictures, notably by Roelant Savery, are skillful and detailed oil paintings and provide a great deal of information as to the appearance of the bird. About a dozen complete skeletons and many isolated bones have been recovered from Mauritius in the past hundred years. Most of the illustrations indicate an extremely clumsy, fat, usually dark gray or blackish bird with a huge beak, minute functionless wings, and a short plumelike tail. It has been supposed that the fatter specimens with less conspicuous tails, darker plumage, and brown rather than golden irides are females, though no contemporary observer noted any sexual dimorphism. Some drawings show a curiously gaunt and perhaps featherless bird. Oudemans supposes that this appearance was assumed seasonally, presumably after breeding, when the time of the main food supply was past. Hachisuka thinks that part of the sheath of the beak was also shed at the molt. The contemporary accounts indicate that the bird lived in forests, fed on fallen fruits of some sort, and laid one white egg about 10 cm long in a nest on the ground.[4]

Most ornithologists have followed Strickland and Melville in regarding the *Raphidae* as aberrant pigeons, the peculiar *Didunculus* of Samoa being supposed to provide a model of how a pigeon started to become a dodo or a solitaire. A few authorities have dissented from their view, and quite recently Lüttschwager[5] has concluded that these strange birds were more closely related to the rails. This view is partly based on details of osteology, but it gains some support from what little is known of the life history, ecology, and behavior of the birds. The recently rediscovered New Zealand *Notornis*, the largest surviving flightless rail, has been found to lay a

4. M. C. Latimer, "A Dodo Egg," *S. Afr. J. Sci.*, *49* (1953), 208–09, According to J. Vinson, *Proc. Soc. Arts Sci. Mauritius*, *1* (1956), 313–15, Latimer's alleged dodo egg, which passed to him from a Miss Lavinia Bean, who obtained it from a sea captain, who received it as a settlement of a debt from the descendant of a sailor shipwrecked on Mauritius, is extremely doubtful.

5. J. Lüttschwager, "Zur systematischen Stellung der ausgestorbenen Drontevögel Gattung *Raphus* (*Didus*) und *Pezophaps*," *Zool. Anz.*, *162* (1959), 127–48.

single egg, which disposes of the difficulty of putting the *Raphidae* in a group which is ordinarily rather fecund. The territorial behavior of the solitaire discussed below seems to Lüttschwager far more rail-like than pigeon-like.

It is just possible that some of the fossil bones will be found to contain enough protein for an immunological test of the possible affinities to be undertaken in the future, though this obviously should not be done on such relatively rare material until the technical details of the procedure are rather better developed than they appear to be at the moment. Part of the difficulty in making judgments of affinities is that birds, though so numerous and superficially diverse, are fundamentally all rather closely related; the anatomical differences between rails and pigeons are by no means great.

The raphid bird of Rodriguez, known generally today as the solitaire, or scientifically *Pezophaps solitaria* (Gmelin), is in some ways the best known of the group, because it was the only species that came under the eye of anyone who was really interested in observing it. Unfortunately, this gentleman, M. François Leguat, who went to Rodriguez in 1691 at the head of the Huguenot settlers, was by his own admission an inexpert draughtsman and so only published one detailed delineation of the bird from life. This is the whole of the iconographic evidence for its appearance, but the general form and some details of his figure have been confirmed by a great quantity of skeletal material recovered during the last century. There is, however, a special complication to be faced in considering the solitaire, not present in the case of the birds of the other islands. It has been seriously proposed that, unlike his bird, Leguat himself is a mythical creation.

In the *Voyage et avantures de François Leguat & de ses Compagnons en deux isles désertes des Indes Orientales,*[6] the solitaire is described as taller than a turkey with similar feet and beak, though the latter was a little more hooked. The male is described as grayish and brown,

6. Two vols., London, David Mortier, Marchand Libraire, 1708. The subsequently published English translation is inaccurate and misleading.

the female as either blonde or brown; one may perhaps suspect an age difference here. The wings were too small for flight; in the account of the male it is said that there was a bony mass under the feathers as big as a musket ball, used in defense. The female had "*une espèce de bandeau comme un bandeau de veuves en haut du bec qui est de couleur tanée.*" It is thus clear from the original, though not from contemporary translations, that the beak was tan-colored. The bandlike structure is clearly dark and conspicuous in Leguat's figure. The most striking feature of the female, however, was that at the base of the neck there were two elevations, with whiter feathers than the rest, "*qui représente merveilleusement un beau sein de femme.*" These structures, which must have put Leguat in mind of some of the ladies he had seen dressed in the fashions of late seventeenth-century France, evidently contributed to his idea that "*la femelle est d'une beauté admirable,*" though the blonde plumage was doubtless most attractive. Of such birds walking to meet him with so proud a display he writes, "*On ne peut s'empêcher de les admirer & de les aimer, de sorte que souvent leur bonne mine leur a sauvé la vie.*" This statement from the first settler on an uninhabited island, in the presence of an edible bird—"*le goût en est excellent sur tout quand ils sont jeunes*"—is a tribute to the elegance of the bird and the sensitivity of the observer.

Leguat continues with an account of the territorial behavior of these birds. The breeding territory had a radius of about two hundred yards, and was defended during incubation and the long period of several months while the young bird was helpless. Leguat says that males drove off intruding males, and females, intruding females. This is reasonable, but it is less likely that the male called the female when he saw an intruding female and vice versa, as Leguat believed that he had observed several times. The wing is said to have been rotated, making a noise like a rattle when the birds were calling. This was presumably a territorial warning. The nest consisted of a heap of palm leaves; a single egg, bigger than that of a goose, was laid; both sexes incubated. Except for what is said about the one sex calling its mate to drive off an intruder of the opposite sex, this is all reasonable, though in Leguat's time it would have seemed less reasonable than it does to us today. Leguat is clearly one of the founders of our knowledge of territoriality in animals, as Armstrong[7] pointed out independently just before the original publication of this essay.

Leguat's description of the end of the rearing of the young, and of the formation of mated pairs, which he says were stable during life, is much odder. Some days after the young bird left the nest, a company of thirty or forty brought another young bird and, the parents joining the band, the two young birds were led off to an unoccupied territory and left there by the adults. He refers to this as a "marriage," and goes on to moralize on the simple fidelity of these birds in a state of nature without laws, lawyers, or priests. This observation of the marriage is apparently unique in avian ethology; Leguat's credibility is therefore a matter of considerable interest.

Leguat's book has had a most curious history. Written in French, it was first published in that language in England in 1708. A number of editions in French, English, and Dutch followed in the same year and the work evidently caused considerable excitement. Right from the beginning doubts as to the authenticity of the work were raised, and though an obituary notice exists, no other independent evidence of Leguat's reality was known until fairly recently. The book was for a long time the only important source for the existence and nature of the Rodriguez solitaire and was accepted as such by Strickland, though by 1848 bones of the solitaire had already been discovered. When later complete skeletons of *Pezophaps* became known, Leguat's remarks about the "*petite masse ronde comme une balle de mousquet*" under the wing feathers received confirmation. In view of this skeletal verification, Alfred Newton and others felt that the work had been vindicated and a reprint was published by the Hakluyt Society in 1891. Practically nothing has been added to

7. E. A. Armstrong, "Territory and Birds: A Concept which Originated from Study of an Extinct Species," *Discovery, 14* (1953), 223–24.

our knowledge of the solitaire until recently, though the osteological evidence on the one hand and Leguat's account on the other have been usually considered as providing a fairly good understanding of the bird.

In a series of works in the 1920s an American scholar, G. Atkinson,[8] undertook to demolish the whole of Leguat's work, claiming that it was a novel written by F. M. Misson. In Atkinson's attack on Leguat, the unhappy solitaire was dealt many undeserved blows.

Atkinson claimed that "Leguat," that is Misson, had invented the solitaire, using as his sources the two brief descriptions by Carré and Du Bois of a bird of the same name in Réunion; a description of the "*oiseau de Nazareth*," apparently a dodo, seen on Mauritius by Cauche; other notes on Mascarene birds given by Cauche and Du Quesne; and finally, for the peculiar skeletal features of the wing, some experience with the skeletons of swans and other European species. The remark, for instance, about the bandeau on the tan-colored beak is supposedly derived from Cauche: "*Leurs perdris ont le bec rouge, il y en a de tannées.*" The practiced philologist will doubtless see the inevitability of the comparison; the present writer does not. To put the solitaire together in a convincing and, in part, osteologically confirmed form, would be a precognitive feat placing Misson as the most successfully imaginative paleontologist who has ever existed. Since Atkinson wrote, the researches of Vivielle and Dehérain have shown that there is independent evidence for the reality of Leguat's voyage in the maritime archives of Texel, Cape Town, and Mauritius, and in some contemporary Dutch naval correspondence; Mortensen has shown that some of Atkinson's geographical and biological criticisms are invalid. Nevertheless, in spite of all this, there is still a tendency to cast doubt on Leguat's testimony. Hachisuka's book reviews, if a little unsystematically, the whole problem, and comes out strongly in Leguat's favor. The

8. *The Extraordinary Voyage in French Literature from 1700 to 1720* (Paris, Lib. Ancienne Honoré Champion, 1922). This gives a summary of the author's views.

territorial behavior recorded, which Atkinson dismisses most unceremoniously, seems to the present writer greatly to strengthen the case for Leguat's reliability, since there were practically no sources available as hints of such behavior in 1708. Perhaps the marriage of the solitaires also really occurred.

A very peculiar aspect of the skeleton of the solitaire is that almost all individuals seem to have suffered from a pathological condition of the bones, which were very easily fractured and produced exostoses. Ottow[9] who has investigated the matter concludes that the condition was comparable to osteogenesis dysplastico-exostotica in man, and was probably inherited. The "*petite masse ronde*" under the wing feathers of the male seems to have been a more or less standardized exostosis, but it is conceivable that as it was developed by natural selection, in the absence of predators or competitors, the whole skeleton became pathological. A very small change in the environment might then have led to extinction not directly at the hand of man. The bird seems to have disappeared during the eighteenth century, but under what circumstances is unknown.

The identity of the bird on Réunion raises a problem which has still not been solved. There are four contemporary descriptions. Tatton, who was in Réunion in 1613, writes of "a great fowl of the bigness of a Turkie, very fat, and so short winged that they cannot flie, beeing white and in a manner tame." Bontekoe was in the island in 1618 and observed "*eenige dod-eersen*" unable to fly and very fat. Carré in 1668 writes of "*celuy que les habitans ont nommé l'Oiseau Solitaire, parce qu'effectivement il aime la solitude . . . Il ne ressembleriot pas mal in Coq d'Inde, s'il n'avoit point les jambes plus hautes. La beauté de son plumage fait plaisir à voir. C'est une couleur changeante qui tire sur le jaune. La chair en est exquise.*"

In the next year Sieur D. B. (Du Bois) wrote, "Solitaires: *ces oiseaux sont nommés ainsi, parce qu'ils vont toujours seuls. Ils sont gros*

9. R. Ottow, "Die erbbedingte Osteogenesis Dysplastico-exostotica der ausgerotteten flugunfähigen Reisentauben *Pezophaps solitaria*, der Mascareneninsel Rodriguez," *K. Svensk. Vetensk. Akad. Handl.*, ser. 4, *1* (1950), no. 9.

comme une grosse Oye, et ont le plumage blanc, noir a l'extremité des ailes et de la queue. A la queue il y a des plumes approchantes de celles d'Autruche, ils ont le col long, et le bec fait comme celui de Bécasses, mais plus gros, les jambes et pieds comme poulets d'Inde. Cet oiseau se prend à la course, ne volant que bien peu."

Strickland,[10] who collated these accounts for the first time, dismissed Bontekoe's use of the term "*dod-eerse*," the ordinary Dutch for dodo, as a confusion with the Mauritius bird, due perhaps to muddled memory caused by Bontekoe's ship being blown up and he alone surviving, shortly after visiting Réunion. Strickland concludes that the three other accounts refer to a yellow or whitish solitaire-like bird with a longer beak than the Rodriguez bird.

It is very curious that Carré's and Du Bois' accounts, which have really confused the issue if only a single species was present, both began in a way reminiscent of the standard opening of any chapter in a medieval bestiary. One can easily imagine a fabulous species being introduced with the words *Avis solitaria vocatur quia solitaria est . . .*

Some years after Strickland wrote, four very definite paintings of white dodos, by Peter Withoos and by Pieter Holsteyn, Sr., were identified in various European collections. It became evident that a dodo with whitish plumage, very decomposed, yellowish wings and plumose tail, actually existed. It was clearly distinct from the Mauritius dodo in its color. It has been supposed that the paintings represented an albino Mauritius dodo. If, however, the bill was less hooked, as seems likely, presumably the bird depicted could only have come from Réunion. Some authorities, notably Rothschild,

10. These extracts are all taken from Strickland. The scientific name of the birds of Réunion raises difficulties. De Selys-Longchamp named the solitaire of Réunion, on the basis of Carré and Du Bois, *Apterornis solitaria.* The generic name is preoccupied, but if a single species existed on the island it becomes *Raphus solitarius* (De Selys-Longchamp) if a dodo, or *Ornithaptera solitaria* if generic distinction is desired. If two birds are involved, the last is the name for the solitaire; Hachisuka has supplied *Victoriornis imperialis* Hachisuka for the white dodo.

have therefore suspected the presence of two white forms on Réunion—one dodo-like—after the paintings, and perhaps Tatton and Bontekoe; the other solitaire-like after the inadequate accounts of Carré and Du Bois. Others, notably Oudemans, suppose only one species to have existed, and this a white dodo. Oudemans, however, thought that the male was deeper colored and was depicted in certain paintings that are now referred by Hachisuka to young Mauritius dodos.

During the present century, at least three good representations have turned up that appear to Hachisuka to represent a solitaire-like bird, either white or brown, with a very large head and conical beak. These he considers to be Réunion solitaires, and, perhaps, overgeneralizing Leguat's rather specialized concept of feminine beauty, he depicts in his reconstruction the female as the more beautiful white bird, largely from a water-color drawing discovered by Killermann in Vienna. It is at first sight hard to assimilate this drawing to a dodo, but if the white dodo lost weight and the anterior sheath of its beak during the molting season, such an identity might be possible.[11] The decomposed wing of the Vienna drawing is very like that of the paintings of white dodos, and it seems inconceivable that two raphid birds, whitish, with the same wings, should have evolved in parallel on the same island. If the white dodos painted by Withoos and by Holsteyn were albinos from Mauritius, then the Réunion bird was presumably solitaire-like. The long neck and beak like a woodcock seem to suggest such an identification of Du Bois' bird. One minor piece of evidence, that it was exquisite to Mons. Carré's palate, also fits in better with the solitaire than the dodo.

The name solitaire may well, as Atkinson implies, have just been

11. Miss Martha M. Dimock, in copying the pictures reproduced in Plate 2, came to the conclusion that the two left-hand paler birds might well be of the same species. This sort of testimony is of some interest in default of direct observation, because in moving the hand in accordance with the old drawing one probably gets nearer to the original bird than in any other way now possible.

used for whatever inhabited Réunion. The habits of the Rodriguez bird as recorded by Leguat are not really in accord with the name. Only skeletal evidence can now solve the problem, and unfortunately the volcanic caves on the island have not yielded bones. The matter is of considerable general interest in view of the difficulties that might arise in explaining the origin of two sympatric raphid birds on the same island. These difficulties are, moreover, presented in a more acute form by the probable former presence of three species of giant tortoise on Rodriguez and of even more on Mauritius. A re-examination of the available material of these animals using modern statistical methods might be interesting.

Finally, it may be mentioned that Peter Mundy in the seventeenth century wrote of the flightless birds of the Mascarenes and of a flightless rail that he, but no one else, observed alive on Ascension Island: "A question may bee Demaunded how they should bee here and not elsewhere, beeing soe Farre From other land and can neither Fly nor Swymme; whether by mixture off kindes producing straunge and monstrous Formes, or the nature off the Climate, ayre and earth in alltring the First Shapes in long time, or how."[12] When another English traveler asked a like question of himself after visiting the Galapagos Islands nearly two centuries later the answer proved to change the ideas of the whole world.

12. "Peter Mundy's Travels," *Hakluyt Soc. Ser.*, *2*, no. 46 (1919), pp. 352–53, and no. 78 (1936), p. 83.

THE USES OF BEETLES[1]

Beetles serve for divers uses, for they both profit our mindes, and they cure some infirmities of our bodies.

A natural history museum, considered in terms of its purposes, may be defined as an institute for the study of the diversity of nature. This is true even if the student is a four-year-old child who can learn that although monkeys and elephants are very different, in each both front and back legs are organized on the same principle, "one bone, two bones and a lot of little bones." It is equally true of the paleontologist using the same principle of homology in attempting to solve the still obscure problem of the origin of the five-toed limb. It was such diversity in unity that Charles Darwin perceived when he compared his skins of Galapagos mockingbirds from the different islands and found that their beaks differed, writing in his notebook in 1835, "When I see these Islands in sight of each other and possessed of but a scanty stock of animals tenanted by these birds but slightly differing in structure and filling the same place in nature, I must suspect they are only varieties. . . . If there is the slightest foundation for these remarks, the Zoology of Archipelagoes will be well worth examination; for such facts would undermine the stability

1. Contribution to a symposium on "The Role of the Museum in Teaching and Research at Yale," held by the Yale Alumni Board on the occasion of the opening of the Oceanographic and Ornithological Wing of the Peabody Museum.

of species." This was the germ of his interest in evolution, the full significance of which we have doubtless not yet appreciated.

Several ancient peoples, notably the Egyptians and the Aztecs, kept a variety of wild animals for pleasure. Aristotle, the most significant pure biologist of the ancient world, is said to have received collections that Alexander the Great sent back from his campaign to his old teacher. Little, however, can be said about the history of natural history in the ancient world. The immediate precursors of the collections in the buildings, the opening of which we are attending, do not go back much beyond four hundred years. This four hundred years of history is, however, very illuminating.

The history of natural history in the Western World may be subsumed under three periods. The first or medieval, from the twelfth century to the early sixteenth century, saw the beginning of the study of Greek and Arabic texts, and the production of a few original works such as the biological writings of Albertus Magnus and the famous treatise of Frederick II on falconry, the relation of which to earlier Arabic works apparently requires further study. The main significance of the medieval period does not lie in these sporadic literary contributions, which were sparse in comparison with the ever-popular and almost entirely fabulous bestiaries based on the "Physiologus." What is really important in the medieval period is the development of an iconography of animals and plants by sculptors and illuminators. We cannot talk about the origin of species, though many people have of course tried, until we know what we mean by species. The sculptors who correctly distinguished, in the Chapterhouse at Southwell, between the two British oaks, were laying the foundation of this knowledge. Somewhat later, illuminators often introduced plants, insects, and birds into the decorative borders of manuscripts. The most astonishing of such decorations, full of insects, spiders, and shells, with a quite recognizable carpet beetle, have been studied by A.C. Crombie of Oxford.[2]

2. "Cybo d'Hyères: A Fourteenth-century Zoological Artist," *Endeavour*, *11* (1952), 183–87.

They date from the fourteenth century and are ascribed to Cybo d'Hyères, a Genoese who seems to have worked in Provence, in a region which was undoubtedly somewhat more influenced by Gothic natural iconography than was much of Italy. There are numerous later examples; a particularly lovely one, a late fifteenth-century "Book of Hours," formerly at Aldenham House and recently from the library of Yale's great benefactor Louis Rabinowitz, is in the Rare Book Room of the Sterling Library. In this exquisite work, believed to have been executed in Ghent or Bruges and certainly dating from the late fifteenth century, the flowers and fruits, which doubtless kept still, are all naturalistic; the birds, which did not, are mainly the types that later turn up on painted china or lacquered screens, though a hoopoe, a goldfinch, and a great-tit are clearly recognizable. The butterflies, of intermediate activity, are fairly realistic, although the artist responsible had some ideas of his own. He depicted a clouded yellow (*Colias croceus*), the European ally of our sulphur butterfly, on several pages (Plate 3, *above*), a wall butterfly (*Pararge megera*) on folio 44v and folio 96r, and an Apollo butterfly (*Parnassius apollo*) on folio 108r (Plate 3, *middle*); but there are also some strangely intermediate forms and at least one blue and yellow species (folios 54r and 54v) whose habitat can only have been the artist's imagination. The presence of the Apollo butterfly is very interesting, for it is an Alpine form that could not have occurred near Ghent or any other Flemish center of the illuminator's art, save as a very rare straggler. It suggests that a set of illustrations of butterflies for the use of illuminators and other decorators may have been circulated. This is probably more likely than the transfer of actual and rather fragile specimens. At least one ornithological manuscript[3] surviving from the fourteenth century probably

3. In the Pepysian Library, Magdalene College, Cambridge, and reproduced in part in *Illustrated London News, 235* (Dec. 12, 1959), pt. 4, facing p. 853. In the caption to this colored reproduction, David Lack points out that one of the birds illustrated is a red-legged partridge, not native to Britain, though there is evidence from a few bird names that the manuscript was made or at least annotated in England.

LATE 3. *Above: Colias croceus*, deep yellow with dark borders to wings; *left*, contemporary specimen (× 2/3), *ight*, miniature Aldenham-Rabinowitz "Book of Hours" folio 100r (× 2).
Aiddle: Parnassius apollo, white with black markings and crimson ocellus on hind wings; *left*, contemporary pecimen (× 2/3), *right*, miniature from Aldenham-Rabinowitz "Book of Hours" folio 108r (× 2).
elow, left to right: Corixa punctata, C. macrocephala, and *C. affinis* (all × 2 and all British specimens).

represented such a collection of bird drawings. Along with illustrated herbals, which have a good Greek ancestry, such treasuries of illustrations of birds and perhaps insects may have been the forebears of the monographs all systematists now use.

Within a hundred years of this work, we have clear evidence that several people were collecting plants and animals, writing about them, drawing them, even attempting to publish about them. The second period in Western natural history had begun. The origin of such collections poses certain interesting problems on which too little research has been done.

The storage of dry drugs in comparably shaped but diversely labeled pots in an apothecary's shop gives some of the diversity in unity that satisfies a collector, as do the spices in neat boxes in a kitchen. The great collecting craze of the Middle Ages was for relics. At least in countries touched by the Reformation, the collection of both works of art and, later, of natural curiosities may have taken the place of the vast accumulations of minute fragments of a whole army of saints, neatly labeled, authenticated, and stored in glazed ornamental containers. A further stimulus to collecting doubtless came from natural rarities such as narwhal's tusks, doing duty for unicorns' horns, and bezoar stones, which were collectors' items, as well as more or less magical drugs.

Sometime during the sixteenth century, the esthetic interest in organisms, the practical interest in drugs, the learned interest in the writings of the ancients that dealt with natural history, and the collectors' enthusiasm previously directed toward books, relics, and *objets d'art* must have united to produce the first systematic natural history collections. Gesner was making collections of plants before 1560, Samuel Quickelberg published a catalogue of his collections in 1565, and there was a storehouse of natural rarities at the court of Saxony. William Penny, who died in 1588, had a "treasury of insects" and left a mass of manuscripts, which later were edited by Thomas Moffet as the "Theatrum Insectorum," from which the epigraph at the beginning of this chapter is taken.

The material brought back from the distant parts of the world that became known from the late fifteenth century onward added greatly to the interest of these and subsequent collectors. It is, moreover, not impossible that knowledge of the diversity of the fauna of Mexico, in which region species after species had a native name, stimulated a systematic attitude in the minds of anyone who had seen what was published of Hernandez' great work on Mexican natural history.[4]

It is important to notice that throughout the whole of the second or Renaissance period of natural history, which lasted from sometime in the sixteenth century to 1859, there is always a strong religious motive for the study. "He that beholds the forms, clothing, elegancy and rich habits of the Butterflies, how can he choose but admire the bountiful God, who is the Author and giver of so rich treasure." In a like vein, Linnaeus believed that his "Systema Naturae" represented valid insight into the thoughts of the Creator. This attitude persisted so long as people believed that they were created to praise God.

The third or modern period of natural history has lasted just a century. It has been dominated by the idea of evolution, a scientific idea that has had a far more powerful effect on men's minds than have any of the physical sciences, however much they may have altered our material environment. Evolution as we know it was born in the mind of Charles Darwin when collecting museum specimens. Much of the later development of the theory has been due to museum workers. A large part of the justification of any university museum lies in the contribution that can be made in it to this kind of knowledge. We need to know far more than we do.

4. Francisco Hernandez, *Nova plantarum, animalium et mineralium mexicanorum historia a Francisco Hernandez Medico in Indiis praestantissimo primum compilata, dein a Nardo Antonio Reccho in volumen digesta, a Io. Terentio, Io. Fabro, et Fabio columna Lynceis notis, & additionibus longe doctissimis illustrata* (Rome, Sumptibus Blasii Deversini, & Zunobii Masotti Bibliopolarum, Typis Vitalis Mascardi, 1651).

Discoveries of any sort in science can be applied to human affairs, but it is evident that in general one needs to know vastly more in order to make the applications wisely and constructively than is needed to do it foolishly and destructively. That is one cause of our present predicament. We are presumably still evolving; people makes guesses that we are undergoing evolutionary loss of teeth, hair, or little toes. For most such guesses, there is not a particle of evidence. There is, however, much evidence to show that, for example, the peculiar characteristic of having a big toe shorter than the second toe, a character that apparently is slightly maladaptive, was regarded as esthetically ideal by the Greeks, and is known to be inherited, has spread through a large part of the population. No one knows why. We are full, body and mind, of such problems. In most cases, we can learn best about ourselves when guided by analogies from birds, fishes, butterflies, sea shells, and sponges.

There is yet another cause of our present discontent. The theory of evolution has often been applied unwisely, sometimes wickedly, to men's minds. A hundred years ago it became evident that each species was not made in Eden individually on the basis of some archetypal design kept in Heaven. When such an oversimple idea was abandoned, the religious meaning for the study of natural history, with much of the wonder and glory, officially disappeared. I do not think that this fundamental attitude really disappeared from the mind of the investigator; a love of the created world, a sort of natural piety, exists in the minds of nearly all naturalists. What happened was that the attitude became something to be held in private, in an apologetic manner, something regarded in the public world as sentimental, impractical, or unrealistic. I believe very strongly that, whatever religious beliefs an investigator may or may not have, this point of view is an essential ingredient of good work, and in our human predicament can be a saving grace. It is one of the attitudes that can help us from destroying our natural heritage, which we are doing at an alarming and ever-increasing rate. I would remind you that during nearly all the history of our species man has lived in association with large, often terrifying, but always exciting animals. Models of the survivors, toy elephants, giraffes and pandas, are an integral part of contemporary childhood. If all these animals became extinct, as is quite possible, are we sure that some irreparable harm to our psychological development would not be done? In a broader context, are we certain that we can really persist as humane human beings in a world in which natural and much artificial beauty is being continually replaced by ugliness or at best by neutral functional forms?

A few weeks ago, I happened to be in the Coe Memorial Room at the top of the new building where our enthusiastic ornithologists were arranging the collection of bird skins. Suddenly it dawned on me that I had never realized what an extraordinary number of pigeons are bright green. Most of you will also probably not have experienced any large number of green pigeons, though to the ornithologist they are a commonplace. Many of them have in addition minor decoration in a great variety of other colors, often of a rather startling kind. To me, this realization, though it had no apparent value in relation to anything else that I knew, gave me intense pleasure that I can still recall and re-experience. Feelings of this sort mold our lives. I think always enriching them. It is for the further opportunity for such experience that I would like to thank all who are responsible for the splendid addition that is being formally presented to the University today.

FIFTY YEARS OF MAN IN THE ZOO

Thirty-six years ago the writer was invited to dine with Professor Punnett. The guest of honor was Thomas Hunt Morgan, the most celebrated biologist that America had produced. Punnett had just confirmed, characteristically working with sweet peas, the main results of the chromosomal theory of inheritance, elaborated over the two previous decades by Morgan, Bridges, and Sturtevant from the results of their studies of the genetics of the pomace fly *Drosophila*. Equally characteristically Punnett had a superb cellar. At the last moment a Russian expert on artificial insemination, then applied mainly to horses, was included in the party. When asked by this Russian guest if there would be opportunities for practicing his art in the New World, Morgan, evidently misunderstanding the true meaning of Punnett's claret, unwisely replied, "Only if you can apply it to Ford cars." The machine age had come upon us. Perhaps *Drosophila* could teach us everything we needed to know about life, and it lived conveniently in rows of cream bottles in a thermostat. Today the same attitude is exhibited by those who look to bacteria and viruses to provide the universal intellectual stimulus of the living world.

It is the purpose of the present chapter to show, without too much nostalgic wearing of the Old Lascaux tie, that this point of view, which grew up during the past half century, has also proved itself, during that time, to be wrong. By considering man as an animal, we may see how essential is the animal kingdom, in its incredible diversity, to a proper understanding of man.

Man possesses bones, and we may conveniently begin by putting him in perspective as a fossilizable animal.

The last million or two years of the earth's history were characterized by the development of great ice sheets over parts of the temperate regions. These ice sheets waxed and waned four to six times, and at least some of the mild interglacial periods were a little warmer than the present. The evolution of our own species *Homo sapiens* from its immediate precursors evidently took place during this dramatic time, which is geologically termed the Pleistocene.

In 1911 a number of fossil skeletons of more or less modern types of man were available, including the famous Cromagnon people who have achieved a reputation, perhaps a little undeserved, for physical perfection and brain size. There were also known some good skeletons of the related beetle-browed *Homo neanderthalensis*, the true caveman of cartoons and comic strips. A huge jaw from Mauer near Heidelberg had been described as a third species of *Homo*, *H. heidelbergensis*. Outside Europe, Dubois, following an extraordinary intuition, had discovered at Trinil in Java the skullcap and femur on which he based the famous *Pithecanthropus erectus*, the first real ape-man, but he became so worried by his discovery that in 1911 it was still not properly described. Just after this time, in 1913, Dawson's skillful artifacts from Piltdown, which later did so much to impede a clear understanding of human evolution, became available. None of these finds could be dated absolutely in years, and in many cases there was great doubt as to the relative antiquity of the different specimens.

The modern period of paleoanthropology really began in the 1920s when the great series of bones of the Peking ape-man, now known as *Pithecanthropus erectus pekinensis*, began to appear from the cave at Chou-koutien, and when Young discovered, and Dart described as *Australopithecus africanus*, an infant skull from Taungs in South Africa. Since then a seemingly endless series of fossils have been found in Asia and Africa. As a result of Leakey's brilliant investigation of Olduvai Gorge in Tanganyika, new material is turning up so fast that anything written at the moment is likely to be out of date by the time it is printed. We also have available, from the method of radiocarbon dating, a reliable chronology for the last 50,000 years of our history, and the promise of the more recent potassium-argon technique becoming applicable to the entire Pleistocene and beyond.

What we know is very briefly as follows. At the beginning of the Pleistocene, about one and three quarters million years ago, there lived in Africa a series of what have been called, rather misleadingly, man-apes, scientifically termed the *Australopithecinae*, who walked erect, and some of whom made simple chopperlike stone tools. The tools found in the lowest bed at Olduvai are believed to be the oldest human artifacts known from anywhere in the world today, and the fossil australopithecines (*Zinjanthropus* and the more recently discovered, slightly older, unnamed fossils) may therefore be regarded as fairly close to the extremely low-browed ancestors of the reader and writer. Several species of australopithecines lingered on much later in South Africa; all were small brained, comparable to modern anthropoid apes in cranial capacity, and there has been much controversy as to their toolmaking capacities.

Much later at Olduvai, typical early Stone Age hand axes appear, and these are associated with what Leakey regards as the first true fossil man of the genus *Homo*. Whether he arose from *Zinjanthropus* on the ground is as yet uncertain.

In Asia there now appears the ape-man related to Dubois' *Pithecanthropus*. As well as the Peking form, today considered as only a subspecies of *P. erectus*, there are also a number of Javanese fossils. *Pithecanthropus* was bigger brained than the australopithecines, but in certain minor anatomical details seems to be off the main line of human descent and probably represents a side branch that came to nothing, The great mandible from near Heidelberg and certain North African fossils have been supposed to be related Western forms.

We then find in the middle Pleistocene in Europe, from Steinheim in Germany, Fontéchevade in France, and Swanscombe in England, a group of imperfect skulls, to which Leakey has compared his Olduvai *Homo*, that are quite modern in character, though with thicker heavier bone that we ourselves possess. From some such stock two lines seems to have developed; one became the beetle-browed caveman *H. neanderthalensis*, the other gave rise to the modern types. Both species or subspecies clearly had developed an elaborate culture with religious beliefs early in their history, for both practiced burial. When portraiture first appears, the first portrait sculpture, from an open last-glacial site in Czechoslovakia, seems to depict an intellectual young woman, who today might be doing brilliantly in graduate school. She is clearly European in facial characters. The various subspecies of *Homo sapiens*, whose existence gives rise to so many practical problems today, seem therefore to have arisen before the last retreat of the ice, probably in response to the very varied climates that man was beginning to experience at this time.

A few rather curious hints about evolution in the Pleistocene can be gathered from certain physiological and anatomical facts. Darwin knew that man is the only animal that blushes. Goodhart has pointed out recently that blushing is a fairly complex physiological process known in all living men, though useless as a signal if the skin is black or if it were covered with hair. It is therefore probable that all existing human populations had a relatively light-skinned and fairly hairless common ancestor. The detailed distribution of such hair as is retained on the human body does not correspond to the pattern of well-developed hair tufts in any other anthropoid. It is not unreasonable to suppose that the deeply pigmented skins of some African and other peoples are secondary adaptations. Leakey has made the astonishing discovery in the lowest Olduvai bed of what seems to be a tool for scraping skins. Modern great apes spontaneously use large leaves as hats, and young chimpanzees in captivity take great delight in wearing human clothing. These facts taken in conjunction with Goodhart's ideas suggest, at least as an interesting speculation, that clothing, with all its immense psychological significance, may have first been made, though not with a needle, before man's brain had reached its full size.

When we try to go further back in time the record is extremely disappointing for some tens of millions of years. During the geological period known as the Miocene, ten to twenty million years ago, there lived in Africa an assemblage of primitive apes or monkeys, placed in the genus *Proconsul*, which seem fairly well qualified as human ancestors. They were clearly animals of the open country, though no doubt on occasion they climbed trees. Their gait was certainly quadripedal; the hand had not yet been fully freed by its possessor's getting up on the hind feet. Among their numerous primate contemporaries they would not have appeared as particularly remarkable. They had generalized simian teeth with moderate canines, the molars arranged in two parallel rows and not in the neat parabolic line that the australopithecines share with ourselves. They lacked the simian shelf, a ridge of bone that strengthens the mandibles of the modern great apes and which makes stripping bark with their great canines a less jawbreaking occupation. In the fine technical detail of the tooth cusps, *Proconsul* suggests a human ancestor, and most investigators think it must have been close to the human line. Leakey has discovered two jaws of apes with simian shelves in a deposit which was evidently laid down in water bordered by forests, rather than in open country. These jaws strongly suggest that at the time of *Proconsul* the line leading to the modern great apes was already distinct; these two fossils in fact are all that we have to elucidate the descent of the chimpanzee, a much more obscure matter than the descent of man.

In time and in structure there is unhappily very little yet to show how a form like *Proconsul* developed into the *Australopithecinae*. It is in this region that at the moment most links are missing. We do have in *Oreopithecus*, from the intervening period in Italy, an arboreal animal with an astonishingly human jaw and pelvis; we also have two fragments from India which also suggest the australopithecine jaw. *Oreopithecus* has a habit of getting into American magazines and newspapers, but it is almost certainly in its hominid characters a strange parallel development. The Indian form *Ramapithecus* looks promising; there is not enough of it to provide any serious difficulties, save that it did not live in Africa.

Before *Proconsul* the problem consists of elucidating the whole history of the early primates. There are dozens of fragmentary fossils representing what must have been charming animals in life. In a general way they show us, as their survivors among the lower primates also do, the gradual process of the eyes' turning forward that permitted binocular vision, the focusing of attention on a single object, a necessary prerequisite for the development of the hand and the mind.

Dry bones can certainly be made to live, if one has an enormous amount of material, great knowledge, enough love for the subject, and a lively but controlled imagination. The process is, however, difficult and the results still rather meager. To get further we must turn to studies of behavior. At least four major approaches have been developed in the past fifty years, though all have their roots in the late nineteenth century. They are the study of brain physiology, of academic experimental psychology, of psychiatry, and of the kind of scientific natural history that is nowadays called ethology. Unfortunately, until recently many of the proponents of the various schools of psychological study devoted too much energy to attacking each other rather than getting on with the problems to be solved. Curiously enough, perhaps the greatest unifying force in recent years has come from what was initially an entirely different field of study. The use of self-regulatory machines had developed early in the nineteenth century but reached an extraordinary development during World War II. In this development new and complicated kinds of applied mathematics had to be invented. Since a missile guided by photocells to seek a target is behaving very much like a man guided by his eyes, walking across a room and putting his hand on a door knob, it is not surprising to find that some of the theory developed in one case applies to the other. In particular it is known deductively that if too great demand is put on any self-regulatory system it is apt to go into oscillation, and that the best way to avoid this is to have a hierarchy of regulatory devices. This arrangement is exactly what the brain physiologists and neurological surgeons have been exploring, while the psychiatrists have had to deal with all sorts of cycloid conditions that look like cases of too great demand being placed on a self-regulatory system. The academic psychologist, in his attempt to establish quantitative laws of the way behavior is learned, has come to realize that the simplicity of his systems is only apparent. A higher organism, even a laboratory white rat, giving simple responses to simple stimuli does so largely because at every stage from perception to motor response an intricate system of controls is operating to keep the system stable and therefore dependable in everyday situations. This point of view is a healthy reminder that we must never say glibly that something is nothing but something else that is supposed to be simpler.

It is very difficult to summarize the work on behavior relative to our theme; there is so much of it but it is still far too compartmentalized. What is clear is that insofar as immediate emotions can be recognized in objectively recordable behavior, man differs very little from most of his primate relations. Some are more social than he is, some less, some exhibit aggression more, some less easily than man does, but as far as one can tell there is not a great deal of difference. Everyone who has worked at all extensively with chimpanzees feels that in all nonintellectual ways they are beings like ourselves, and the same is true with increasing limitations as we go from the higher to the lower primates. Recent events have tended to make people skeptical of *Homo sapiens* as the rational animal that the schoolmen and the *philosophes* believed him to be, but it is quite certain that he is much more intellectual than any other animal and that in no other important way is he unique.

The evolution of the hand and of speech certainly are physical prerequisites of intellectual ability but without the latter they would merely be curious adaptations comparable to hundreds of elaborate specializations found up and down the animal kingdom. It is becoming clear that all large mammals capable of learning are potentially able to develop dispositions and capacities quite different from what they ordinarily develop in nature. Elsa, in Joy Adamson's wonderful book, became a civilized lioness; she had clear limitations and remained a lioness, but was unlike any lioness that had ever lived before. A single unfavorable review of *Born Free* dismissed the book as dealing with the least pressing problem of the twentieth century. It is possible that it dealt with one case, and a very illuminating case, of a class of problems which includes all the really pressing issues of our time.

In considering both man's evolution and his behavior we inevitably face another related field where zoological analogies have proved fruitful in illuminating a region of the utmost practical importance. Man is an animal that exhibits a quite extraordinary tendency to what zoologists call subspeciation, the tendency to produce locally distinct populations. These are usually referred to as races, but that word is so charged with emotions that the more neutral term subspecies is preferable in the present context. The phenomenon occurs in most widely distributed animals; an ornithologist can easily tell a robin that has bred in Newfoundland from those of Connecticut, Virginia, or California. It is, however, probably more strongly developed in man than in any other species, doubtless on account of his great geographical range. Originally in man, as in other organisms, the subspecific characters were probably partly adaptations to climate; stocky individuals with short extremities in cold regions conserved heat, while in hot countries long spindly legs help to dissipate heat. All the circumstances that promoted these adaptations can now be taken care of by artifacts more effectively. The morphological adaptations have largely lost their meaning, and probably started to do so more than 10,000 years ago.

There has been much movement of populations since then, so that it is now hard to trace the original meaning of the variations. A black skin may once have been an advantage in hunting by night. Now it is a disadvantage on an ill-lit road traveled by automobiles. Unhappily this is not the only disadvantage of being heavily pigmented in North America.

Since, for about four-fifths (say 40,000 out of 50,000 years) of their duration as subspecific entities, all men were living by their wits in the Stone Age, the amount of wits needed was probably everywhere much the same. Therefore, if we look without prejudice, we find no evidence of differences in *average* innate intellectual ability among the principal human subspecies. What we find, and nothing would lead us to suspect this a priori, is that within populations there is an enormous individual amount of variation. Man is highly polymorphic, to use the technical term, in intellectual ability. Until recently this fact, though obviously of great sociological importance, has never been taken very seriously. In order to learn more about it, other cases of polymorphism must be studied. The classical cases are in the color patterns of butterflies and moths, and it is these cases which provided the initial suggestions in studying human polymorphism. With this acknowledgment to the Lepidoptera, we can consider first some human cases.

The best known ones concern blood groups, because survival after transfusion appeals to people more than intellectual distinction and appalls them less than congenital feeblemindedness. The most interesting cases involved are the peculiar conditions of the red blood corpuscles known as sickling and thalassaemia. These conditions occur mainly where certain forms of malaria are or have been prevalent. They are inherited in such a way that if an individual receives a sickling gene from each parent he dies early of a serious form of anemia, if from one parent he has functionally normal blood but with red corpuscles that, when collapsed in certain solutions, have a characteristic sickle or half-moon shape. An individual receiving the sickling gene from neither parent is, of course, in almost every case normal with respect to the condition. The evolutionary question that arises is why, if the individuals receiving the sickling gene from both parents always die, the condition does not disappear from the population, as Darwin would lead us to think. The answer appears to be that the blood corpuscles of the hybrids with one sickling gene, though functional, are just abnormal enough chemically to provide bad homes for certain stages of the malaria parasites, to which, therefore, the hybrid individuals are immune. Normal individuals therefore get malaria, hybrids do not, people with two sickling genes get anemia; the hybrids are therefore the fittest that survive. When they mate, according to Mendel's law, they produce on an average one normal prone to malaria, two hybrids, and one anemic. The polymorphism therefore persists. There is more than a suggestion that most of the multifarious inherited blood types behave in a comparable though much less extreme way.

The question now arises as to whether psychological diversity can be due to the same sort of mechanism. Though the expression of intelligence is certainly enormously influenced by upbringing, morons and intellectuals appear in the most unexpected places. Moreover, the woefully inadequate work on the characteristics of identical twins reared apart shows clearly that there is an inherited influence in determining at least the scores in intelligence tests. Penrose has found that among the Scots, who are traditionally committed to encouraging intellectual capacities, the population of any generation tends to be recruited from the people of middle ability in the previous generation. The moron and the genius contribute less than the *homme moyen sensuel*. Here then we seem to have a polymorphic situation of the greatest social significance, comparable to the sickling gene case but doubtless dependent on many pairs of genes, which goes far to explain the distribution actually observed in all large human populations. If we introduced social mechanisms to encourage the production of more intellectuals, perhaps society might perish from a dearth of stupidity.

There is one final zoological case which shows how possibly temperamental and emotional differences might be in part of genetic origin but in a very unobvious way. King and Shea, in a recent but already classical study, have examined the basis for an inherited difference in capacity to climb in two subspecies of a single species of deer mouse. The open-country subspecies opens its eyes a little earlier in its career than does the woodland form. If the two kinds of mice are reared in the laboratory with no opportunity for vertical movement till they are relatively mature, both forms climb. If, however, the opportunity to climb is presented from the beginning, the open-country form tries to climb as soon as it can see, but is inadequate to the task, falls down and learns that climbing is a bad thing to do. The woodland form, developing vision and muscular coordination somewhat later, climbs well from the beginning and goes on doing so. If we were deer mice and the tendency to climb had great social significance, while ethical considerations prevented direct experimentation, we should have two rival schools, environmentalists who claimed that the difference was due to upbringing, which it is, and geneticists who claimed that it was due to heredity, as indeed it is. There would be much emotion involved; the situation would be typically human. It is more than a guess that this sort of thing does indeed occur in man.

As yet it is not possible to summarize all the diverse themes of this essay in a concise and formal set of statements. Yet there is one conclusion, as to method if not results, that is of great importance. The themes that have been developed have been illuminated by studies on butterflies, deer mice, robins, a lioness, and by implication all the primates living as well as fossil. It would have been quite possible to develop other themes involving snails, waterbugs, birds of paradise, dolphins, giraffes, and rhinoceroses. We are only at the beginning of this kind of study. A world containing none of these animals would be intellectually impoverished to a degree that we are only just beginning to realize. We are, however, rapidly approaching such a world. Most Western countries, the U.S.S.R., and Japan are becoming alive to the danger. Unfortunately a great many of the new countries, where a vast number of the most interesting animals live, regard them as a nuisance to be rid of, and interest in them as a symbol of colonialism. There are encouraging signs that the shortsightedness of such ideas are beginning to be realized. In view of the desire of the United States, both altruistic and practical, to help these countries, it is of great importance that those concerned in such aid should understand their responsibility in helping to maintain an adequate diversity of living beings on our planet, for without this diversity it will be immeasurably more difficult to understand ourselves.

Note in proof: It now seems possible that *Zinjanthropus* did not make the crude stone tools associated with his remains, and that a more advanced form existed contemporaneously. These results of Leakey's latest discoveries have so far appeared mainly in reports of lectures.

HOMAGE TO SANTA ROSALIA

or

WHY ARE THERE SO MANY KINDS OF ANIMALS?[1]

When you did me the honor of asking me to fill your presidential chair, I accepted perhaps without duly considering the duties of the president of a society founded largely to further the study of evolution, particularly at the close of the year that marks the centenary of Darwin and Wallace's initial presentation of the theory of natural selection. It seemed to me that most of the significant aspects of modern evolutionary theory have come either from geneticists, or from those heroic museum workers who, suffering through years of neglect, were able to establish about twenty years ago what has come to be called the "new systematics." You had, however, chosen an ecologist as your president and one of that school at times supposed to study the environment without any relation to the organism.

A few months later I happened to be in Sicily. An early interest in zoogeography and in aquatic insects led me to attempt to collect, near Palermo, certain species of waterbugs, of the genus *Corixa*, described a century ago by Fieber and supposed to occur in the

1. Address of the President, American Society of Naturalists, delivered at the annual meeting, Washington, D.C., December 30, 1958. Dr. A. Minganti of the University of Palermo enabled me to collect on Monte Pellegrino. Bernard M. Knox of the Department of Classics of Yale University gave me a rare and elegant word from the Greek to express the blurring of a food chain. L. B. Slobodkin of the University of Michigan and R. H. MacArthur of the University of Pennsylvania provided me with their customary kinds of intellectual stimulation. To all these friends I am most grateful.

region, but never fully reinvestigated. It is hard to find suitable localities in so highly cultivated a landscape as the Concha d'Oro. Fortunately, I was driven up Monte Pellegrino, the hill that rises to the west of the city, to admire the view. A little below the summit, a church with a simple baroque façade stands in front of a cave in the limestone of the hill. Here in the sixteenth century a stalactite-encrusted skeleton associated with a cross and twelve beads was discovered. Of this skeleton nothing is certainly known save that it is that of Santa Rosalia, a saint of whom little is reliably reported save that she seems to have lived in the twelfth century, that her skeleton was found in this cave, and that she has been the chief patroness of Palermo ever since. Other limestone caverns on Monte Pellegrino had yielded bones of extinct Pleistocene *Equus*, and on the walls of one of the rock shelters at the bottom of the hill there are beautiful Gravettian engravings. Moreover, a small relic of the saint that I saw in the treasury of the Cathedral of Monreale has a venerable and petrified appearance, as might be expected. Nothing in her history being known to the contrary, perhaps for the moment we may take Santa Rosalia as the patroness of evolutionary studies, for just below the sanctuary, fed no doubt by the water that percolates through the limestone cracks of the mountain and which formed the sacred cave, lies a small artificial pond, and when I could visit the pond a few weeks later, I got from it a hint of what I was looking for.

Vast numbers of *Corixidae* were living in the water. At first I was rather disappointed because every specimen of the larger of the two species present was a female and so lacking in most critical diagnostic features, while both sexes of the second, slightly smaller species were present in about equal number. Examination of the material at leisure, and of the relevant literature, has convinced me that the two species are the common European *C. punctata* and *C. affinis*, and that the peculiar Mediterranean species are illusionary. The larger *C. punctata* was clearly at the end of its breeding season; the smaller *C. affinis* was probably just beginning to breed. This is the sort of observation that any naturalist can and does make all the time. It was not until I asked myself why the larger species should breed first, and then the more general question as to why there should be two and not twenty or two hundred species of the genus in the pond, that ideas suitable to present on this occasion began to emerge. These ideas finally prompted the very general question as to why there are such an enormous number of animal species.

There are at the present time supposed to be about one million described species of animals.[2] Of these about three-quarters are insects, of which a quite disproportionately large number are members of a single order, the Coleoptera.[3] The marine fauna, although it has at its disposal a much greater area than has the terrestrial, lacks this astonishing diversity.[4] If the insects are excluded, it would seem to be more diverse. The proper answer to my initial question would be to develop a theory at least predicting an order of magnitude for the number of species of 10^6 rather than 10^8 or 10^4. This I certainly cannot do. At most it is merely possible to point out some of the factors which would have to be considered if such a theory were ever to be constructed.

Before developing my ideas I should like to say that I subscribe to the view that the process of natural selection, coupled with isola-

2. S. W. Muller, and A. Campbell, "The Relative Number of Living and Fossil Species of Animals," *Systematic Zoology*, *3* (1954), 168–70. L. H. Hyman, "How Many Species?" *Systematic Zoology*, *3* (1955), 142–43.
3. There is a story, possibly apocryphal, of the distinguished British biologist J. B. S. Haldane, who found himself in the company of a group of theologians. On being asked what one could conclude as to the nature of the Creator from a study of his creation, Haldane is said to have answered, "An inordinate fondness for beetles." However, the late Hugh Scott, who had an unrivalled knowledge of all order of insects, wrote just before his death, in comment on the original version of this footnote, that he believed that when the group was better known the Diptera might rival the Coleoptera in numbers and that the proper response might well have been, "An inordinate fondness for flies."
4. G. Thorson, "Bottom communities," chap. 17 in *Treatise on marine ecology and paleoecology*, Vol. 1., *Geol. Soc. Amer. Memoir*, *67* (1957), 461–534.

tion and later mutual invasion of ranges leads to the evolution of sympatric species, which at equilibrium occupy distinct niches, according to the Volterra-Gause principle. The empirical reasons for adopting this view, and the correlative view that the boundaries of realized niches are set by competition, are mainly indirect. So far as niches may be defined in terms of food, the subject has been carefully considered by Lack.[5] In general all the indirect evidence is in accord with the view, which has the advantage of confirming theoretical expectation. Most of the opinions that have been held to the contrary appear to be due to misunderstandings and to loose formulation of the problem.[6] The relatively rare exceptions involve special complications, mainly symbiosis, that can be treated by changes of sign or other devices in the mathematical development of the deductive theory.

In any study of evolutionary ecology, food relations appear as one of the most important aspects of the system of animate nature. There is quite obviously much more to living communities than the raw dictum "eat or be eaten," but in order to understand the higher intricacies of any ecological system, it is most easy to start from this crudely simple point of view.

Animal ecologists frequently think in terms of food chains, of the form: *individuals of species S_1 are eaten by those of S_2, of S_2 by S_3, of S_3 by S_4*, etc. In such a food chain S_1 will ordinarily be some holophytic organism or material derived from such organisms. The simplest case is that in which we have a true *predator chain* in Odum's convenient terminology,[7] in which the lowest link is a green plant, the next a herbivorous animal, the next a primary carnivore, the next a secondary carnivore, etc. A specially important type of

5. D. Lack, *The Natural Regulation of Animal Numbers* (Oxford, Clarendon Press, 1954).
6. G. E. Hutchinson, "Concluding remarks," *Cold Spring Harbor Symp. Quant. Biol., 22* (1958), 415–27.
7. E. P. Odum, *Fundamentals of Ecology* (Philadelphia and London, W. B. Saunders, 1953).

predator chain may be designated Eltonian, because in recent years C. S. Elton[8] has emphasized its widespread significance, in which the predator at each level is larger and rarer than its prey. This phenomenon was recognized much earlier, notably by A. R. Wallace in his contribution to the 1858 communication to the Linnean Society of London.[9]

In such a system we can make a theoretical guess of the order of magnitude of the diversity that a single food chain can introduce into a community. If we assume that in general 20 per cent of the energy passing through one link can enter the next link in the chain, which is overgenerous,[10] and if we suppose that each predator has twice the mass (or 1.26 the linear dimensions) of its prey, which is a very low estimate of the size difference between links, the fifth animal link will have a population of one ten-thousandth (10^{-4}) of the first, and the fiftieth animal link, if there was one, a population of 10^{-49} the size of the first. Five animal links are certainly possible, a few fairly clear-cut cases having been in fact recorded. If, however, we wanted 50 links, starting with a protozoan or rotifer feeding on algae with a density of 10^6 cells per ml, we should need a volume of 10^{26} cubic kilometers to accommodate on an average one specimen of the ultimate predator, and this is vastly greater than the volume of the world ocean. Clearly the Eltonian food chain of itself cannot give any great diversity, and the same is almost certainly true of the other types of food chain, based on detritus feeding or on parasitism.

Before proceeding to a further consideration of diversity, it is,

8. C. S. Elton, *Animal Ecology* (London, Sedgwick and Jackson, 1927).

9. A. R. Wallace, "On the tendency of varieties to depart indefinitely from the original type," *in* C. Darwin and A. R. Wallace, *On the Tendency of Species to Form Varieties; and on the Perpetuation of Varieties and Species by Natural Means of Selection*. Originally published in *J. Linn. Soc. London* (*Zool.*), *3* (1858), 45–62.

10. R. L. Lindeman, "The Trophic Dynamic Aspect of Ecology," *Ecology*, *23* (1942), 399–408. Cf. also Slobodkin, *Amer. Nat.*, *94* (1960), 213–36, who gives 15 per cent as a reasonable upper limit.

however, desirable to consider the kinds of selective force that may operate on a food chain, for this may limit the possible diversity.

It is reasonably certain that natural selection will tend to maintain the efficiency of transfer from one level to another at a maximum. Any increase in the predatory efficiency of the nth link of a simple food chain will, however, always increase the possibility of the extermination of the $(n - 1)^{th}$ link. If this occurs either the species constituting the nth link must adapt itself to eating the $(n - 2)^{th}$ link or itself become extinct. The process will in fact tend to shortening of food chains. A lengthening can presumably occur most simply by the development of a new terminal carnivore link, as its niche is by definition previously empty. In most cases this is not likely to be easy. The evolution of the whalebone whales, which, at least in the case of *Balaenoptera borealis*, can feed largely[11] on copepods and so rank on occasions as primary carnivores, presumably constitutes the most dramatic example of the shortening of a food chain. Mechanical considerations would have prevented the evolution of a larger rarer predator, until man developed essentially non-Eltonian methods of hunting whales.

A second important limitation of the length of a food chain is due to the fact that ordinarily animals change their size during free life. If the terminal member of a chain were a fish that grew from say 1 cm to 150 cms in the course of an ordinary life, this size change would set a limit by competition to the possible number of otherwise conceivable links in the 1–150 cm range. At least in fishes, this type of process (metaphoetesis) may involve the smaller specimens of any given species belonging to links below the larger, and the chain length is thus lengthened, though under strong limitations, by cannibalism.

We may next inquire into what determines the number of food chains in a community. In part the answer is clear, though if we cease to be zoologists and become biologists, the answer begs the

11. H. B. Bigelow, "Plankton of the Offshore Waters of the Gulf of Maine," *Bull. U.S. Bur. Fisheries*, 40 (1926), 1–509.

question. Within certain limits, the number of kinds of primary producers is certainly involved, because many herbivorous animals are somewhat eclectic in their tastes and many more limited by their size or by such structural adaptations for feeding that they have been able to develop.

The extraordinary diversity of the terrestrial fauna, which is much greater than that of the marine fauna, is clearly due largely to the diversity provided by terrestrial plants. This diversity is actually twofold. Firstly, since terrestrial plants compete for light, they have tended to evolve into structures growing into a gaseous medium of negligible buoyancy. This has led to the formation of specialized supporting, photosynthetic, and reproductive structures which inevitably differ in chemical and physical properties. The ancient Danes and Irish are supposed to have eaten elm bark, and sometimes sawdust, in periods of stress, has been hydrolyzed to produce edible carbohydrate; but usually man, the most omnivorous of all animals, has avoided almost all parts of trees except fruits as sources of food, though various individual species of animals can deal with practically every tissue of many arboreal species. A major source of terrestrial diversity was thus introduced by the evolution of almost 200,000 species of flowering plants, and the three-quarters of a million insects supposedly known today are in part a product of that diversity. But of itself merely providing five or ten kinds of food of different consistencies and compositions does not get us much further than the five or ten links of an Eltonian pyramid. On the whole the problem still remains, but in the new form: why are there so many kinds of plants? As a zoologist, I do not want to attack that question directly; I want to stick with animals, but also to get the answer. Since, however, the plants are part of the general system of communities, any sufficiently abstract properties of such communities are likely to be relevant to plants as well as to herbivores and carnivores. It is therefore by being somewhat abstract, though with concrete zoological details as examples, that I intend to proceed.

Biological communities do not consist of independent food chains, but of food webs, of such a kind that an individual at any level (corresponding to a link in a single chain) can use some but not all of the food provided by species in the levels below it.

It has long been realized that the presence of two species at any level, either of which can be eaten by a predator at a level above, but which may differ in palatability, ease of capture, or seasonal and local abundance, may provide alternative foods for the predator. The predator, therefore, will neither become extinct itself nor exterminate its usual prey, when for any reason, not dependent on prey-predator relationships, the usual prey happens to be abnormally scarce. This aspect of complicated food webs has been stressed by many ecologists.[12] Recently MacArthur,[13] using an ingenious but simple application of information theory, has generalized the points of view of earlier workers by providing a formal proof of the increase in stability of a community as the number of links in its food web increases.

MacArthur concludes that in the evolution of a natural community two partly antagonistic processes are occurring. More efficient species will replace less efficient species, but more stable communities will outlast less stable communities. In the process of community formation, the entry of a new species may involve one of three possibilities. It may completely displace an old species. This of itself does not necessarily change the stability, though it may do so if the new species inherently has a more stable population than the old. Secondly, it may occupy an unfilled niche, which may, by providing new partially independent links, increase stability.

12. Particularly W. C. Allee, A. E. Emerson, O. Park, T. Park, and K. P. Schmidt, *Principles of Animal Ecology* (Philadelphia and London, W. B. Saunders, 1949); Odum, *loc. cit.*; and C. S. Elton, *The Ecology of Invasions by Animals and Plants* (London, Methuen, 1958).

13. R. H. MacArthur, "Fluctuations of Animal Populations and a Measure of Community Stability," *Ecology*, *35* (1955), 533–36. Cf also L. B. Slobodkin, "Conditions for Population Equilibrium," *Ecology*, *35* (1955), 530–33.

Thirdly, it may partition a niche with a pre-existing species. Elton,[14] in a fascinating work largely devoted to the fate of species accidentally or purposefully introduced by man, concludes that in very diverse communities such introductions are difficult. Early in the history of a community we may suppose many niches will be empty and invasion will proceed easily; as the community becomes more diversified, the process will be progressively more difficult. Sometimes an extremely successful invader may oust a species but add little or nothing to stability, at other times the invader by some specialization will be able to compete successfully for the marginal parts of a niche. In all cases it is probable that invasion is most likely when one or more species happen to be fluctuating and are underrepresented at a given moment. As the communities build up, these opportunities will get progressively rarer. In this way a complex community containing some highly specialized species is constructed asymptotically.

Modern ecological theory therefore appears to answer our initial question at least partially by saying that there is a great diversity of organisms because communities of many diversified organisms are better able to persist than are communities of fewer less-diversified organisms. Even though the entry of an invader which takes over part of a niche will lead to the reduction in the *average* population of the species originally present, it will also lead to an increase in stability, reducing the risk of the original population being at times underrepresented to a dangerous degree. Thus, loss of some niche space may be compensated by reduction in the amplitude of fluctuations in a way that can be advantageous to both species. The process, however, appears likely to be asymptotic, and we have now to consider what sets the asymptote, or in simpler words, why are there not more different kinds of animals?

It is first obvious that the processes of evolution of communities must be under various sorts of external control, and that in some

14. *Loc. cit.*

cases such control limits the possible diversity. Several investigators[15] have pointed out that the more or less cyclical oscillations observed in arctic and boreal fauna may be due in part to the communities' not being sufficiently complex to damp out oscillations. It is certain that the fauna of any such region is qualitatively poorer than that of warm temperate and tropical areas of comparable effective precipitation. It is probably considered to be intuitively obvious that this should be so, but on analysis the obviousness tends to disappear. If we can have one or two species of a large family adapted to the rigors of arctic existence, why can we not have more? It is reasonable to suppose that the total biomass may be involved. If the fundamental productivity of an area is limited by a short growing season to such a degree that the total biomass is less than under more favorable conditions, then the rarer species in a community may be so rare that they do not exist. It is also probable that certain absolute limitations on growth-forms of plants, such as those that make the development of forest impossible above a certain latitude, may in so acting severely limit the number of niches. Robert MacArthur points out that the development of high tropical rain forest increases the bird fauna more than that of mammals, and Thorson likewise has shown that the so-called infauna shows no increase of species toward the tropics while the marine epifauna becomes more diversified. The importance of this aspect of the plant or animal substratum, which depends largely on the length of the growing season and other aspects of productivity, is related to that of the environmental mosaic discussed later.

We may also inquire, but at present cannot obtain any likely answer, whether the arctic fauna is not itself too young to have achieved its maximum diversity. Finally, the continual occurrence of catastrophes, as Wynne-Edwards[16] has emphasized, may keep the

15. Odum, *loc. cit.*; MacArthur, *loc. cit.*

16. V. C. Wynne-Edwards, "Zoology of the Baird Expedition (1950): I. The Birds Observed in Central and Southeast Baffin Island," *Auk*, *69* (1952), 353–91.

arctic terrestrial community in a state of perennial though stunted youth.

Closely related to the problems of environmental rigor and stability is the question of the absolute size of the habitat that can be colonized. Over much of Western Europe there are three common species of small voles, namely *Microtus arvalis*, *M. agrestis*, and *Clethrionomys glareolus*. These are sympatric but with somewhat different ecological preferences.

In the smaller islands off Britain and in the English channel[17] there are only two cases of two species' co-occurring on an island, namely *M. agrestis* and *Clethrionomys* on the islands of Mull and Bute, in the Inner Hebrides. On the Orkneys the single species is *M. orcadensis*, which in morphology and cytology is a well-differentiated ally of *M. arvalis*; a comparable animal (*M. sarnius*) occurs in Guernsey. On most of the Scottish Islands only subspecies of *M. agrestis* occur, but on Mull and Raasay, as well as on the Welsh island of Skomer, and on Jersey, races of *Clethrionomys* of somewhat uncertain status are found. No voles have reached Ireland, presumably for paleogeographic reasons, but they are also absent from a number of small islands, notably Alderney and Sark. The last named island must have been as well placed as Guernsey to receive *M. arvalis*. Still stranger is the fact that although it could not have got to the Orkneys without entering the mainland of Britain, no vole of the *arvalis* type now occurs in the latter country. Cases of this sort may be perhaps explained by the lack of favorable refuges in randomly distributed, very unfavorable seasons, or under special kinds of competition. This seems very reasonable as an explanation

17. See particularly G. S. Miller, *Catalogue of the Mammals of Western Europe* (London, British Museum, 1912).
G. E. H. Barrett-Hamilton, and M. A. C. Hinton *A History of British Mammals*, Vol. 2 (London, Gurney and Jackson, 1911–1921).
Lord Cranbrook, "Longtailed Field Mice *Apodemus* from the Channel Islands," *Proc. Zool. Soc. London*, *128* (1957), 597–600.
M. J. Delany, "The Ecological Distribution of Small Mammals in Northwest Scotland, *Proc. Zool. Soc. London*, *137* (1961), 107–26.

of the lack of *Microtus* on Sark, where it may have had difficulty in competing with *Rattus rattus* in a small area. It would be stretching one's credulity to suppose that the area of Great Britain is too small to permit the existence of two sympatric species of *Microtus*, but no other explanation seems to have been proposed.

In the Channel Islands[18] only Jersey has two species of shrew, though there are four or five in Central France, excluding the aquatic *Neomys*. One of the Jersey shrews is *Sorex araneus fretalis*, the other a race of *Crocidura suaeveolens*. The latter occurs by itself on Sark, but Alderney, Guernsey, and Herm have *C. russula*. It is reasonable again to suppose that on most such islands only one species can survive.

It is a matter of considerable interest that Lack,[19] studying the populations of birds on some of these small British islands, concluded that such populations are often unstable, and that the few species present often occupied larger niches than on the mainland in the presence of competitors. Such faunas provide examples of communities held at an early stage in development because there is not enough space for the evolution of a fuller and more stable community.

18. The Earl of Cranbrook, and P. Crowcroft, "The White-toothed Shrews of the Channel Islands," *Ann. Mag. Nat. Hist.*, *1*, ser. 13 (1958), 359–64; "Small Mammals from Herm Island," *J. Linn. Soc. London* (*Zool.*), *44* (1961), 365–68. The distribution of the shrews of the Channel Islands elicited by the Earl of Cranbrook and Crowcroft is very peculiar. It seems almost certain that *C. russula* and *C. suaeveolens* must have entered the region together, and by analogy with Ireland and other older small islands of Britain one would have expected *Sorex minutus* to have also been available as an invader at the same time. *C. suaeveolens* must have got into Britain as it is known as a well-defined subspecies on the Scilly Islands, though not on the British mainland. Corbet (*Nature*, *191*, 1961, 1037–40) thinks all these distributions are due to man and his commerce; the fundamental ecological point, however, remains.
19. D. Lack, "Ecological Features of the Bird Faunas of British Small Islands, *J. Animal Ecol.*, *11* (1942), 9–36. The whole subject of the relation of area to number of species on islands has recently been enormously advanced by E. O. Wilson, *Amer. Nat.*, *95* (1961), 169–93.

The various evolutionary tendencies, notably metaphoetesis, which operate on single food chains must operate equally on the food web, but we also have a new, if comparable, problem as to how much difference between two species at the same level is needed to prevent them from occupying the same niche. Where metric characters are involved we can gain some insight into this extremely important problem by the study of what Brown and Wilson[20] have called *character displacement* or the divergence shown when two partly allopatric species of comparable niche requirements become sympatric in part of their range.

I have collected together a number of cases of mammals and birds that appear to exhibit the phenomenon.[21] These cases involve metric characters related to the trophic apparatus, the length of the culmen in birds and of the skull in mammals appearing to provide appropriate measures. Where the species co-occur, the ratio of the larger to the smaller form varies from 1.1 to 1.4, the mean ratio being 1.28 or roughly 1.3. This latter figure may tentatively be used as an indication of the kind of difference necessary to permit two species to co-occur in different niches but at the same level of a food web.

In the case of the aquatic insects[22] with which I began my address, we have over most of Europe three very closely allied species of *Corixa* (Plate 3, *below*), the largest, *punctata*, being about 116 per cent

20. W. L. Brown and E. O. Wilson, "Character Displacement," *Systematic Zoology*, 5 (1956), 49–64.

21. The technical details are given in *Amer. Nat.*, *93*: 153 in a table omitted from the present version. Klopfer and MacArthur (*Amer. Nat.*, *95*, 1961, 223–26), conclude that in *climatically stable* tropical environments far less difference between competing species is needed than in the temperate regions that I have considered. Consequently many more species can coexist.

22. For the relevant natural history of these insects, see particularly T. T. Macan, "A Contribution to the Study of the Ecology of Corixidae (Hemip.)," *J. Animal Ecol.*, *23* (1954), 115–41.
E. S. Brown, "A Contribution towards an Ecological Survey of the Aquatic and Semi-aquatic Hemiptera-Heteroptera (waterbugs) of the British Isles, etc.," *Trans. Soc. British Entom.*, *9* (1958), 151–95.

longer than the middle-sized species *macrocephala*, and 146 per cent longer than the small species *affinis*. In northwestern Europe there is a fourth species, *C. dentipes*, as large as *C. punctata* and very similar in appearance. A single observation suggests that this is what I have elsewhere termed a fugitive species, maintaining itself in the face of competition mainly on account of greater mobility in the southern part of its range, but extending further north than any of the other three species. According to Macan while both *affinis* and *macrocephala* may occur with *punctata* they never are found with each other, so that all three species never occur together. In the eastern part of the range, *macrocephala* drops out, and *punctata* appears to have a discontinuous distribution, being recorded as far east as Simla, but not in southern Persia or Kashmir, where *affinis* occurs. In these eastern localities, where it occurs by itself, *affinis* is larger and darker than in the west, and superficially looks like the middle-sized species *macrocephala*.[23]

This case is very interesting because it looks as though character displacement is occurring but that the size differences among the three species are just not great enough to allow them all to co-occur in the same pond. Other characters than size are in fact clearly involved in the separation, *macrocephala* preferring deeper water than *affinis* and the latter being more tolerant of brackish conditions. It is also interesting because it calls attention to a marked difference that must occur between hemimetabolous insects with annual life cycles involving relatively long growth periods, and birds or mammals in which the period of growth in length is short and of a very special nature compared with the total life span. In the latter, niche separation may be possible merely through genetic size differences, while in a pair of animals like *C. punctata* and *C. affinis* we need not only a size difference but a seasonal one in reproduction; this is likely to be a rather complicated matter. For the larger of two species always to be larger, it must never breed later than the smaller

23. G. E. Hutchinson, "Copepodology for the Ornithologist," *Ecology, 32* (1951), 571–77.

one. I do not doubt that this is what was happening in the pond on Monte Pellegrino, but have no idea how the difference in timing is achieved.

I want to emphasize the complexity of the adaptation necessary on the part of two species inhabiting adjacent niches in a given biotope, as it probably underlies a phenomenon which to some has appeared rather puzzling. MacArthur[24] has shown that in a sufficiently large bird fauna, in a uniform, undisturbed habitat, areas occupied by the different species appear to correspond to the random nonoverlapping fractionation of a plane or volume. Kohn has found the same thing for the cone shells (*Conus*) on the Hawaiian reefs. This type of arrangement almost certainly implies such individual and unpredictable complexities in the determination of the niche boundaries, and so of the actual areas colonized, that in any over-all view the process would appear random. It is fairly obvious that in different types of community the divisibility of niches will differ and so the degree of diversity that can be achieved. The fine details of the process have not been adequately investigated, though many data must already exist that could be organized to throw light on the problem.

A final aspect of the limitation of possible diversity and one that perhaps is of greatest importance, concerns what may be called the mosaic nature of the environment. Except perhaps in open water when only uniform quasi-horizontal surfaces are considered, every area colonized by organisms has some local diversity. The significance of such local diversity depends very largely on the size of the organisms under consideration. Elsewhere MacArthur and I[25] have

24. R. H. MacArthur, "On the Relative Abundance of Bird Species," *Proc. Nat. Acad. Sci. Wash.*, *43* (1957), 293–95; "On the Relative Abundance of Species, *Amer. Nat.*, *94* (1960), 25–36.
A. J. Kohn, "The Ecology of Conus in Hawaii," *Ecol. Monogr.*, *29* (1959), 47–90.

25. G. E. Hutchinson, and R. H. MacArthur, "A Theoretical Ecological Model of Size Distributions Among Species of Animals, *Amer. Nat.*, *93* (1959), 117–26.

attempted a theoretical formulation of this property of living communities and have pointed out that even if we consider only the herbivorous level or only one of the carnivorous levels, there are likely, above a certain lower limit of size, to be more species of small- or medium-sized organisms than of large organisms. It is difficult to go much beyond crude qualitative impressions in testing this hypothesis, but we find that for mammal faunas, which contain such diverse organisms that they may well be regarded as models of whole faunas, there is a definite hint of the kind of theoretical distribution that we deduce. In qualitative terms the phenomenon can be exemplified by any of the larger species of ungulates that may require a number of different kinds of terrain within their home ranges, any one of which types of terrain might be the habitat of some small species. Most of the genera or even subfamilies of very large terrestrial animals contain only one or two sympatric species. In this connection I cannot refrain from pointing out the immense scientific importance of obtaining a really full insight into the ecology of the large mammals of Africa while they can still be studied under natural conditions. It is indeed quite possible that the results of studies on these wonderful animals would in long-range, though purely practical, terms pay for the establishment of greater reservations and national parks than at present exist.

In the passerine birds the occurrence of five or six closely related sympatric species is a commonplace. In the mammal fauna of Western Europe no genus appears to contain more than four strictly sympatric species. In Britain this number is not reached even by *Mustela* with three species. On the adjacent parts of the continent there may be three sympatric shrews of the genus *Crocidura*, and in parts of Holland three of *Microtus*. In the same general region there are genera of insects containing hundreds of species, as in *Atheta* in the Coleoptera and *Dasyhelea* in the Diptera Nematocera. The same phenomenon will be encountered whenever any well-studied fauna is considered. Irrespective of their position in a food chain, small size, by permitting animals to become specialized to the conditions offered by small diversified elements of the environmental mosaic, clearly makes possible a degree of diversity quite unknown among groups of larger organisms.

We may, therefore, conclude that the reason why there are so many species of animals is at least partly because a complex trophic organization of a community is more stable than a simple one, but that limits are set by the tendency of food chains to shorten or become blurred, by unfavorable physical factors, by space, by the fineness of possible subdivision of niches, and by those characters of the environmental mosaic which permit a greater diversity of small than of large allied species.

In conclusion I should like to point out three very general aspects of the sort of process I have described. One speculative approach to evolutionary theory arises from some of these conclusions. Just as adaptive evolution by natural selection is less easy in a small population of a species than in a larger one, because the total pool of genetic variability is inevitably less, so it is probable that a group containing many diversified species will be able to seize new evolutionary opportunities more easily than an undiversified group. There will be some limits to this process. Where large size permits the development of a brain capable of much new learned behavior, the greater plasticity acquired by the individual species will offset the disadvantage of the small number of allied species characteristic of groups of large animals. Early during evolution the main process from the standpoint of community structure was the filling of all the niche space potentially available for producer and decomposer organisms and for herbivorous animals. As the latter, and still more as carnivorous animals began to appear, the persistence of more stable communities would imply splitting of niches previously occupied by single species as the communities became more diverse. As this process continued one would expect the over-all rate of evolution to have increased, as the increasing diversity increased the probability of the existence of species preadapted to new and unusual niches. It is reasonable to suppose that strong predation among macroscopic metazoa did not begin until the late Precambrian, and that the appearance of powerful predators led to the appearance of fossilizable skeletons. Of those advanced so far, this seems the only reasonable hypothesis to account for the relatively sudden appearance of several fossilizable groups in the Lower Cambrian. The process of diversification would, according to this argument, be somewhat autocatakinetic even without the increased stability that it would produce; with the increase in stability it would be still more a self-inducing process, but one, as we have seen, with an upper limit. Part of this upper limit is set by the impossibility of having many sympatric allied species of large animals. These, however, are the animals that can pass from primarily innate to highly modifiable behavior. From an evolutionary point of view, once they have appeared, there is perhaps less need for diversity, though from other points of view, as Elton has stressed in dealing with human activities, the stability provided by diversity can be valuable even to the most adaptable of all large animals. We may perhaps therefore see in the process of evolution an increase in diversity at an increasing rate till the early Paleozoic, by which time the familiar types of community structure were established. There followed then a long period in which various large and finally large-brained species became dominant, and then a period in which man has been reducing diversity by a rapidly increasing tendency to cause extinction of supposedly unwanted species, often in an indiscriminate manner. Finally we may hope for a limited reversal of this process when man becomes aware of the value of diversity, no less in an economic than in an esthetic and scientific sense.

A second and much more metaphysical general point is perhaps worth a moment's discussion. The evolution of biological communities, though each species appears to fend for itself alone, produces integrated aggregates which increase in stability. There is nothing mysterious about this; it follows from mathematical theory and appears to be confirmed to some extent empirically. It is, however, a phenomenon that also finds analogies in other fields in which a more complex type of behavior, which we intuitively regard as higher, emerges as the result of the interaction of less complex types of behavior, which we call lower. The emergence of love as an antidote to aggression, as Lorenz pictures the process, or the development of cooperation from various forms of more or less inevitable group behavior that Allee[26] has stressed, are examples of this from the more complex types of biological systems.

In the ordinary sense of explanation in science, such phenomena are explicable. The types of holistic philosophy that import *ad hoc* mysteries into science whenever such a situation is met are obviously unnecessary. Yet perhaps we may wonder whether the empirical fact that it is the nature of things for this type of explicable emergence to occur is not something that itself requires an explanation. Many objections can be raised to such a view; a friendly organization of biologists could not occur in a universe in which cooperative behavior was impossible, and without your cooperation I could not raise the problem. The question may in fact appear to certain types of philosophers not to be a real one, though I suspect such philosophers, in their desire to demonstrate how often people talk nonsense, may sometimes show less ingenuity than would be desirable in finding some sense in such questions. Even if the answer to such a question were positive, it might not get us very far; to an existentialist, life would have merely provided yet one more problem; students of Whitehead might be made happier, though on the whole the obscurities of that great writer do not seem to generate unhappiness; the religious philosophers would welcome a positive answer but would note that it told them nothing that they did not know before; Marxists might merely say, "I told you so." In spite of this, I suspect that the question is worth raising, and that it could be phrased so as to provide some sort of real dichotomy between alternatives; I therefore raise it knowing that I cannot, and suspecting that at present others cannot, provide an intellectually satisfying answer.

26. W. C. Allee, *Animal Aggregations: A Study in General Sociology* (Chicago Univ. Press, 1931).

My third general point is less metaphysical, but not without interest. If I am right that it is easier to have a greater diversity of small than of large organisms, then the evolutionary process in small organisms will differ somewhat from that of large ones. Wherever we have a great array of allied sympatric species there must be an emphasis on very accurate interspecific mating barriers, which is unnecessary where virtually no sympatric allies occur. We ourselves are large animals in this sense; it would seem very unlikely that the peculiar lability that seems to exist in man, in which even the direction of normal sexual behavior must be learned, could have developed to quite the existing extent if species recognition, involving closely related sympatric congeners, had been necessary. Elsewhere I[27] have attempted to show that the difficulties that *Homo sapiens* has to face in this regard may imply various unsuspected processes in human evolutionary selection. But perhaps Santa Rosalia would find at this point that we are speculating too freely, so for the moment, while under her patronage, I will say no more.

27. See above, pp. 62–76.

MEDITATION ON DEATH IN THE CHURCHES OF FLORENCE

The two great churches of the friars, the Dominican Santa Maria Novella and the Franciscan Santa Croce, are enormous mortuary monuments, as all who study their guidebooks will know. In them almost everyone of importance to the city is commemorated, along with many whose names if decipherable could hardly be called forgotten, for in life they were known to so few.

Visitors go to see the Spanish Chapel in the Cloisters of Santa Maria Novella, where the boat in the vault seems to detach itself from the Sea of Galilee and to float over the rows of Thomistic figures on the walls that so enraged the late Bernard Berenson. To reach the chapel one must pass the lovely medallion (Plate 4) commemorating Cesira Malevolti, "*abilissima in gentili lavori del suo sesso*," who died in her sixteenth year on March 9, 1842. "*Abbandono contenta la terra per rivolare al cielo*"; she is known to us who remain on earth only because she was fortunate in her tomb. Any large Italian cemetery containing enough nineteenth-century sculpture will yield something to the eye of the connoisseur.

Much more rarely have once-living human beings become, involuntarily, their own monuments. Under the steps that go from the left transept to the Strozzi Chapel, where Nardo di Cione depicted Dante's Inferno, Purgatorio, and Paradiso, there lies the incorruptible body of the Blessed Alessio di Jacopo Strozzi, once prior of Santa Maria Novella, who died in 1383. Dried and shrunken though he is, he seems to be otherwise perfect where the skin is

PLATE 4. Cesira Malevolti.

exposed. Corporeal incorruptibility is a very curious problem. The late Father Herbert Thurston, S.J., made a statistical survey[1] of the phenomenon as it affected the bodies of all saints dying between 1400 and 1900 who were of sufficient importance to have their names in the Roman calendar and so a cult of more than local significance. Of such saints there are 42; of them 22, or 52 per cent, exhibited marked evidence of the phenomenon, their bodies remaining supple, and usually emitting a fragrant odor, in some cases for many years after death. The reported cases seem to imply retention of water in the tissues or its replacement by some less volatile liquid material which permits them to remain somewhat soft. The best cases reported from the fifteenth to the eighteenth century seem clearly unlike the numerous more or less mummified bodies so common in relatively modern catacombs in Italy and also different from the bodies in newly opened tombs, often reported in the days before photography, as showing perfectly preserved features that fell to dust on admission of air.

Thurston claims that since his 42 cases were publicly and very widely known for their sanctity in life, or they would not have been entered in the Roman calendar, there is no question of incorruptibility having played any important part in promoting their canonization, and that therefore the sample is unbiased in this respect. Skeptics would no doubt maintain that surreptitious embalming might account for the observed phenomena. There is doubtless much room for subjective interpretation, and no evidence of how frequently the bodies of individuals enforced by poverty to a life of great abstemiousness may show a like tendency in the absence of exceptional sanctity. Of all strange phenomena for which there seems to be some prima facie evidence, this is one of the least known. Some may think that there is little enough here to investigate, others that the investigation would in itself be sacrilegious. Yet 107 years after her death in 1624 in Madrid, the body of the Blessed

1. *The Physical Phenomena of Mysticism*, ed. J. H. Crehan, S. J. (London, Burns Oates, 1952), chap. 10.

Maria Anna Ladroni of Jesus was all but completely dissected by several medical men in an attempt to ascertain the causes of its fresh and soft appearance and of the fragrant oily exudate that delighted the anatomists. "If these proceedings were irregular, irreverent and unwarranted, they were yet so useful in establishing the reality of the miracle that we cannot find it in our heart to condemn them."[2] If only the poor *beata* had been left for a little more than two more centuries, how much could have been learned far more decorously. Unhappily, after the rigorous treatment of 1731, renewed examination showed that what was left of her had become hard and wasted though by no means reduced to dust. There seem to be a few good modern examples, and something might be learned by application of modern techniques to some of the better older cases.

The dismemberment of saints would seem to be a normal fate even if by nature they may exhibit incorruptibility. It is a curious redress of an old and most unnecessary wrong that Galileo should also have been honored in this very ecclesiastical way. His right index finger was cut off when his body was removed to Santa Croce. Still far from being merely a skeleton, it is enshrined in a glass reliquary with gilt fittings, authenticated by a seal in the customary manner, and provided with a poem beginning "*Leipsana ne spernas digiti* . . . ," the first word being also ordinarily applied to the relics of the saints. This now is exhibited with the cracked object glass of the telescope with which he discovered the Galilean satellites of Jupiter, along with the two surviving complete telescopes that he used later, in the fascinating Museo delle Scienze in the old Palazzo de'Giudici.

Whether martyrs' tombs were the first formalized Christian altars has been debated by the learned. At least in the Roman usage, each must now contain relics of four saints, two having suffered martyrdom. In addition to these fragments, much larger relics are often placed beneath altars, from which skulls may seem to stare

2. Quoted by Thurston (pp. 260–62) in translation from Pietro dello Spirito Santo, *Compendio della Vita della B. Marianna di Gesu* (Rome, 1783), p. 126.

between neatly arranged long bones, all carefully labeled. Sometimes a whole skeleton is preserved in this way, but verisimilitude may be achieved by embedding the bones in wax and fashioning a whole new body around the real relics. San Firenze lies now, his own baroque church having been converted into a law court, under a side altar on the right of the nave of San Giovannino degli Scopoli and bears a label, "*Ossa rivestita di cere del giovane S. Florenzio martire sotto Domiziano. . . .*" His waxen androgynous young body wears a short, white, girlish, silk dress, has long black hair with a chaplet of roses, and has many rings on the fingers of his right hand that clasps a silk handkerchief. He ordinarily receives little attention; both the anxious older woman and the somewhat wayward-looking girl are kneeling before the nearby altar of Santa Rita di Cascia, the patroness of hopeless causes and impossible people, on whose feast automobiles are blessed. Where there are no bones, a large figure of a popular saint can be made in wax, or even the dead Christ portrayed with disconcerting realism under an altar.

The early Florentine workers in wax seem mainly to have made votive models of parts of the body to leave at a shrine after a miraculous cure. This is of course a common practice, certainly of pre-Christian origin, though silver is now preferred in most places. During the seventeenth century the dangerous realism of wax came to be fully appreciated, never perhaps more than by the pathological monk and modeler Gaetano Zumbo (1656–1701) from Syracuse, who specialized in the portrayal of the diseased and decomposing dead. During the eighteenth century this art, which had a religious origin and no doubt, whatever its unconscious meaning, was rationalized as a reminder of death, acquired a didactic and scientific purpose in many centers in Italy. In Florence it was carried to an obsessive extreme by Clemente Susini (1754–1814).[3] In the rooms

3. For further information on Susini, see L. Castaldi: "Francesco Boi, 1767–1860, primo cattedratico di anatomia humana a Cagliari e le cere anatomiche fiorentini di Clemente Susini," *Biblioteca della Rivista di Storia delle Scienze Mediche e Naturali*, Vol. 1, 76 pp. (Florence, Olschki, 1947).

of an upper floor of the zoological institute on the Via Romana there is an anatomical museum that houses much of his work, though many models were also exported to centers of medical education elsewhere in Europe. Corpse after corpse modeled with the same realistic fidelity as the figures beneath altars, but each dissected to show some particular organ system, are laid out in the central cases of these rooms. In the side cases are detailed models of special structures, some enlarged. The elaboration of the dissection of the lymphatic system is unbelievable. The whole, or almost the whole, of human anatomy is here. Art has imitated nature to the minutest detail, and though Susini was a professor at the Academia delle Belle Arti, nothing of the spirit of his forebearers remains. The only thing that is not natural is the do-not-touch atmosphere of the place. Here are dozens of incorruptible bodies, but none are supple. Everything that the student needed to know could be learned, but it would be most subtly secondhand. Yet everyone who has taught the structure of man, animal, or plant with a desk model before him owes something to Susini, who must have set a standard never again attained. On leaving the place it is, however, comforting to remember that even later than the date of his death in 1814, the representationalist tradition could still produce something as lovely as Cesira Malevolti.

SIR GEORGE GABRIEL STOKES, 1819–1903

Most names are soon forgotten. Some become impersonal units—watts, ohms, and amperes. A few achieve immortality in connection with major generalizations such as gravitation or evolution. Others have become attached to statements of quantitative laws which may represent only a minute part of the achievement of their authors. It is this somewhat epigrammatic survival which has been the lot of Sir George Gabriel Stokes who died on February 1, 1903, fifty years before these remarks first appeared in print, and who was much concerned theologically with immortality. The demonstration that a very small sphere falling in a viscous medium encounters a resistance proportional to its velocity, its radius, and to the viscosity of the medium is a matter of great importance to the meteorologist, to the industrial physicist, and to the student of plankton, to name three diverse types of investigator. Probably few of the thousands who use this generalization daily know that its discoverer was a professor of mathematics who used physical methods to make fundamental discoveries in biochemistry. Of all the great natural philosophers of the nineteenth century, Stokes is perhaps the least known as a man; it is hoped that these notes will show that his career and character are of interest to the historian of science and to the student of human nature.[1]

1. Most of the personal information, unless otherwise acknowledged, is taken from the memoir by Stokes' daughter, Mrs. Laurence Humphry, *in* Joseph Larmor, *Memoir and Scientific Correspondence of Sir George Gabriel Stokes,*

George Gabriel Stokes was born on August 13, 1819, at Skreen, County Sligo. He came of an English family which had settled in Ireland in the seventeenth century and had produced a number of learned men. His father and both his grandfathers were clergymen of the English Church in Ireland; his paternal great-grandfather, born in 1682, held the office of Deputy Surveyor for Ireland and wrote a treatise on hydrostatics. Today, the position of such a family may appear in retrospect to be somewhat invidious, but this certainly did not occur to the Stokes in the third decade of the nineteenth century. "The disloyal people," wrote Elizabeth Stokes, George's older sister, "were called 'White Boys'; George did not like to go out with his pinafore on, for fear he should be taken for a 'white boy.' "

In the old-fashioned colonial type of society of the Anglo-Irish of that period, the *status quo* was axiomatically correct until proved wrong, and then anything might happen. The resulting peculiar mixture of conservatism and originality clung to Stokes during his entire life. His later career in one of the older English universities, throughout the entire reign of Queen Victoria, was undoubtedly in harmony with such an outlook, but the circumstances of his family background should be kept in mind in judging his life as a whole.

An early proficiency in mathematics impressed his Irish schoolmaster. As a boy, George Stokes appears also to have been a keen

---

Bart. (2 vols. Cambridge Univ. Press, 1907). In addition to Mrs. Laurence Humphry's account, the work also includes biographical appreciations by G. D. Liveing, Sir Michael Foster, Sir Williams Huggins, and Bishop G. F. Browne, colleagues and friends of Stokes during the last third of the nineteenth century. The late Sir Ellis Minns and my mother, the late Mrs. E. D. Hutchinson, of Cambridge, England, also sent me their reminiscences, my father having been a close associate of Stokes during the last years of his life. Stokes' scientific work is easily accessible *in* G. G. Stokes, *Mathematical and Physical Papers*, reprinted from *Original Journals and Transactions* (5 vols. Cambridge Univ. Press, 1880–1905).

field naturalist. His daughter, Mrs. Laurence Humphry, wrote, "One day while he was at Cambridge, returning from a walk, he failed to respond to the salutation of some ladies of his acquaintance; when asked the reason of such odd behavior he answered that he could not bow, as his hat was full of beetles." This incident, so suggestive of a Lear Limerick, is the kind of thing also recorded of Darwin. The full role of the British Coleoptera in the intellectual history of mankind awaits its historian. The elder Rayleigh, writing of Stokes' lectures at a later time, records that the impression left upon the hearer was that Stokes "felt himself as much at home in chemical and botanical questions as in Mathematics and Physics."

Stokes was a strong, well-built man, a great walker and a powerful swimmer. He seems to have delighted in color; his daughter wrote, "My mother and I used to be amused at the vehement colours he wished us to select when we had patterns of dress stuffs. . . . I used to say that he would like us to be dressed in the colours of the spectrum, a different one each time." He indicated that the direction of his interests into hydrodynamics and optics was largely accidental, due to the suggestions of a teacher and the rearrangement of a course of university lectures. There can, however, be little doubt that a direct sensory delight in wind and water, and in the play of light in nature, made these branches of physics peculiarly congenial as fields for the exercise of his extraordinary mathematical skill.

Every account of his personal characteristics mentions that Stokes was a very silent man. Sir Ellis Minns, the great and genial student of the Scythians and the last to survive of Stokes' colleagues at Pembroke College, in some notes he kindly sent me, writes: "As an example of his silence Mrs. A. N. Whitehead told me that for some reason she had to visit in his house about two hours and that they sat on each side of the fire-place and he said never a word." Yet, when he had something to say, he said it well.

As his father died when George Stokes was fifteen, his education was largely planned by his elder brother William, a man of some scientific attainments, interested in mineralogy and chemistry. George Stokes entered Pembroke College in 1837 and was placed first among the men of his year in mathematics in 1841. This success led to his election to a fellowship at his college. After graduation, he continued to reside in Pembroke, taking private pupils as was the custom of the day, and, as he put it, trying his hand at original research. As a result of his achievements, he was elected Lucasian Professor at Cambridge in 1849, so filling from that year, till his death in 1903, the chair which Newton had occupied. Stokes also held the mastership of his college for a few months before his death. He was elected to the Royal Society of London in 1851; the other leading scientific societies of the world followed by conferring foreign membership or associateship on him, including the National Academy of Sciences, Washington, in 1883. He was created a Baronet of the United Kingdom in 1889. Worldly honors[2] meant little to him; he accepted them quietly and with grace, but the documents conferring them seem to have been relegated, along with uncashed dividend checks and other papers of value, to an attic over the stable attached to his daughter's house.

Stokes began publishing the results of his researches in 1842. His purely physical work is mainly concerned with hydrodynamics and optics, with occasional excursions into mechanics; the inevitable Victorian railway bridge appears here. His hydrodynamic studies took him into the field of the figure of the earth and gravitation, and also led him to consider the elasticity of solids, a subject of great optical interest in view of the paradoxical properties postulated for the luminiferous ether. Heat he hardly touched, electricity he left severely alone, these two fields being cultivated by his friends Kelvin and Clerk Maxwell. When Stokes started his career, the great French mathematicians had opened up the field of hydrodynamics, but at Cambridge it was felt that they had sacrificed

2. One peculiar honor may be mentioned; Stokes is commemorated in the mineral name *Stokesite*, a tin metasilicate described from a single good crystal from Cornwall. The mineral was not rediscovered until 1961.

physical reality to mathematical generality. The undulatory theory of light was keenly debated at the time; some of the older men, in spite of the discoveries of Young and of Fresnel, would have nothing to do with it. It was natural for a young investigator of great ability to tackle the difficult field of hydrodynamics, and, when opportunity arose, to plunge headlong into teaching and research in optics.

We are separated from this time by the great triumphs of middle and late nineteenth-century physics, thermodynamics, and the electromagnetic theory of radiation, as well as by everything which came later. The direction of Stokes' researches has caused time to do him an injustice; his finished theoretical studies are embedded impersonally in the standard textbooks. One of his more important mathematical discoveries actually first appeared as a result to be proved in an examination question; although widely known as Stokes' Integral, there seems to be no proper publication by the author. His great experimental discovery of fluorescence received no real explanation until quantum mechanics invaded chemistry, and by then the man who made the original observations was largely forgotten.

Two of the earlier papers may be mentioned as being of particular interest. The first of these is the paper of 1850, "On the Effect of the Internal Friction of Fluids on the Motion of Pendulums." Stokes had become interested in viscosity by 1845, and in April of that year he communicated an important paper, "On the Theories of the Internal Friction of Fluids in Motion, and of the Equilibrium and Motion of Elastic Solids," to the Cambridge Philosophical Society. This paper incidentally indicated the significance of the turbulent wake behind a body moving in water. The original stimulus for the later paper of 1850 seems to have been the investigation by Colonel Sabine in 1829 of the time of vibration of a pendulum in an atmosphere of hydrogen compared with the time in air. These experiments suggested an effect of the gas, independent of density and analogous to the internal friction or viscosity of liquids.

Many investigators, both experimental and mathematical, applied themselves to the subject, but none seemed to have understood the physical nature of the problem. Stokes' interest in it seems to have been aroused partly by the inadequacy of the work of his predecessors, and partly by some observations by Sir James Smith, who, affixing little flags of gold leaf to a pendulum, showed that only at some finite distance away were the tips of the flags bent; the vibrating pendulum was evidently carrying its own atmosphere with it and not sliding through the air in contact with its surface as had been supposed. Stokes provided solutions for a number of possible cases of the general problem. As a special case of his treatment of the ball pendulum, when the velocity is very small and uniform motion replaces oscillatory motion as a limiting case, he obtained the expression[3] for the resistance encountered by a small sphere moving very slowly through a viscous medium:

$$-F = 6\,\pi\,\mu\,a\,V$$

where μ is the viscosity of the medium and a and V the radius and velocity of the sphere. Whence it easily follows that, for a sphere of density p' at its terminal velocity, falling through a medium of density p,

$$V = \tfrac{2}{9}g\,(p' - p)\,\mu^{-1}\,a^2$$

3. Stokes introduced and used throughout this paper the concept of kinematic viscosity or, as he called it, the *index of friction* $\mu' = \mu\rho^{-1}$. The density of the medium, therefore, appears explicitly in his expression for the resistance, which he wrote as $-F = 6\pi\mu'\rho a V$. The simpler expression has been given in the text because it is more familiar and because it appears naturally as a special case, for $n = 1$ and the constant k appropriate for a sphere, of the general expression suggested by the theory of dimensions, namely $-F = k\mu^{2-n}\rho^{n-1}d^nV^n$. When $n = 2$, this reduces to the old expression found by Newton, which is appropriate to a large body falling rapidly, Reynold's number being large and viscosity of no importance. It was such a formulation that misled the early nineteenth-century physicists when they looked at the clouds. The intermediate cases, such as that for $n = 3/2$, investigated by Allen (*Philosophical Magazine*, 50, 1900, 328) are still inadequately studied, being of little importance in meteorology or aerodynamics, though of great biological interest.

Stokes had no very good data for the viscosity of gases. He was in fact later to advance considerably the theory of obtaining such data himself. He was, however, easily able to show from such information as was then available that his new expressions gave much smaller values for the velocity of descent of very small water droplets in air than those given by previously accepted theory. In doing so, he had gone a long way toward explaining the existence of clouds, which, however obvious they may be to the eye, had been inexplicable to earlier natural philosophers.

The second paper which may be mentioned is one "On the Composition and Resolution of Streams of Polarised Light from Different Sources," communicated to the Cambridge Philosophical Society in 1852. This paper has had a most curious fate. The mathematical method which Stokes introduced made little or no impression on his contemporaries, but almost a century later has been revived and clothed in modern dress by Chandrasekhar and Elbert, who have developed it to give an explanation of the remarkable distribution of the polarization of light of the sunlit sky.[4] This result would have delighted Stokes.

It may be of interest to mention, before proceeding to an account of Stokes' greatest experimental discovery, that in 1849 he introduced a practical method of measuring astigmatism, a defect of vision first observed by Airy in his own eyes about twenty years earlier.

In 1845, Sir John Herschel had observed that a dilute solution of quinine sulphate, when illuminated by a beam of white light, showed "an extremely vivid and beautiful celestial blue colour apparently resulting from the action of the strata which the light first penetrates on entering the liquid." Herschel called the phenomenon epipolization, which we know now by Stokes' name, fluorescence. Analogous observations had been made by Sir Daniel Brewster some years earlier, but not published. Herschel was at a loss to

4. S. Chandrasekhar and D. Elbert, "Polarisation of the Sunlit Sky," *Nature*, *167* (1951), 51–55. I am indebted to my friend Rupert Wildt for pointing out the historical significance of this remarkable paper.

explain the phenomenon; Brewster explained it as due to reflection from minute particles, but the fact that the light supposedly so scattered is unpolarized is not in accord with this explanation.

Stokes, attracted by Herschel's observations, examined a quinine solution in a solar spectrum. Moving the solution in a glass vessel through the spectrum, he found that the "rays belonging to the greater part of the visible spectrum passed freely through the fluid, just as if it had been water, being merely reflected here and there from motes. But from a point about half way between the fixed lines G and H (i.e. the Ca lines at 4308 Å and 3969 Å) to far beyond the extreme violet, the incident rays gave rise to a light of a sky blue colour which emanated in all directions from the portion of the fluid which was under the influence of the incident rays. . . . The fixed lines belonging to the violet and invisible region beyond were beautifully represented by dark planes interrupting the blue space." By these observations, first made on Monday, April 28, 1852, Stokes showed that in fluorescence the refrangibility, or, as we should now prefer to say, the wave length, of the light is altered, the fluorescent light in general being less refrangible, or of longer wave length, than the exciting light. "Nothing seems more natural than to suppose that the incident vibrations of the luminiferous ether produce vibratory movements among the ultimate molecules of sensitive substances, and that the molecules in turn swinging on their own account, produce vibrations . . . and thus cause the sensation of light." Stokes had thus discovered a new optical phenomenon which he rightly suspected to be "associated with the inmost structure of chemical molecules to such a degree as to throw even the phenomenon of polarisation into the shade."

Of more immediate importance was the fact that fluorescence provided a new tool for mapping the solar or any other spectrum in the ultraviolet or region of chemical rays. An enormous extension of the ultraviolet spectrum of the electric spark beyond that of the solar spectrum was soon observed. Fluorescence also provided a sensitive series of tests for various organic compounds. One of these, aesculin, in a decoction of horse-chestnut bark, provided a very suitable fluorescent detector. Another convenient detector was a canary-colored glass of Bohemian origin, which Brewster had earlier observed to exhibit a "green internal dispersion," as he put it. Stokes ascertained that this glass contained uranium and set about examining other uranium salts. As is well known, the general fluorescence of such salts provided a stimulus, however illogical it may now seem, to Becquerel in his discovery of radioactive radiation.

Brewster's earlier observations of the internal dispersion of chlorophyll solutions led Stokes to a more elaborate study of chlorophyll pigments. In a short paper published in 1864, he announced that the chlorophyll, that is plastid pigments, of land plants, at first apparently grass cuttings from the lawn of his house, Lensfield Cottage (later Stokes Lea) in Cambridge, consisted of "four substances, two green and two yellow, all possessing highly distinctive optical properties." He goes on to state that green seaweeds "agree with land-plants, except as to the relative proportion of the substances present; but in olive-coloured sea-weeds (*Melanospermeae*) the second green substance is replaced by a third green substance and the first yellow substance by a third yellow substance, to the presence of which the dull colour of these plants is due. The red colouring-matter of the red sea-weeds (*Rhodospermeae*), which the plants contain in addition to chlorophyll, is altogether different in its nature from chlorophyll, as is already known, and would appear to be an albuminous substance. I hope, before long, to present to the Royal Society the details of these researches." But the promised paper never appeared, and Stokes left to others, many years afterward, the development of these fundamental discoveries in plant biochemistry.

About the same time, Stokes was also working on the optical properties of blood. His experiments were done in a passage behind the pantry of his house, and his family complained that his lack of sense of smell, his only physical deficiency, caused him to leave plates of blood around which would have better been thrown away.

The results of these researches were published in a paper in 1864, and led him to conclude that the pigment of blood exists in a reduced and in an oxidized form, which he called purple and scarlet cruorine respectively. He found that "scarlet cruorine is capable of reduction even in the cold by substances present in blood" (Stokes maintained that since the blood did not stink, it was undecomposed; the testimony of his family on his sense of smell is interesting here). By merely shaking the solution with air, the purple cruorine oxidizes to scarlet and "we have every reason to believe that a *portion* of the cruorine present in the venous blood exists in the state of purple cruorine, and is reoxidised in passing through the lungs." Though the history of the discovery of the relationship of oxidized to reduced hemoglobin is a somewhat confused one—Hoppe-Seyler, who introduced the term hemoglobin, having made comparable but not identical observations about the same time—there is no doubt that this paper of Stokes has an honorable place among the many contributions which have completed, on the biochemical side, William Harvey's discovery of the circulation of the blood made two centuries earlier.

The first three volumes of Stokes' collected papers contain memoirs published in 1842–52; the remaining two volumes are sufficient to cover the results of the rest of his scientific life. As did Newton, Stokes had a period of intense productivity, followed by a longer time when he became immersed in public affairs and in theology. Both men were presidents of the Royal Society of London; both men for a time represented their University in Parliament, where neither ever joined in debate.

The causes of Newton's virtual retirement from the field of his genius were in part due to physical ill health, and no doubt were in part psychological and unconscious. Stokes, however, quite possibly emulated his great predecessor consciously. He notes in one of his letters that Newton wrote extensively on religious matters; Stokes' colleagues seem to have used Newton's example in persuading him to stand for Parliament. What Newton did, Stokes deemed appropriate for him to do also.

Stokes accomplished far more scientific work during his later life than the volume of his papers indicates. This work was certainly of less importance than his production from the first hydrodynamic work in 1842 to about 1864, when, at the age of 45, he published his two fundamental biochemical discoveries. The fields he had made his own were at that time pretty well explored by the methods he had cultivated, and could not become fertile again until quite new methods, developed in different branches of science, were again available at the end of the nineteenth century. His studies of the peculiar optical properties of certain potassium chlorate crystals, published in 1885, and various short papers on flames, written six or seven years later, indicate that he was continually experimenting. Any of the phenomena which he studied might have proved as important as the blue color observed by Herschel in his quinine solution, but in fact none did so. Stokes himself realized that in such work there was a continual chance merely of showing "in what precise way the phenomenon was brought about by the operation of well-known causes."

Stokes was also continually helping other workers. He early conceived that the duties of the Lucasian Professor included assisting all members of the University who encountered mathematical obstacles in their researches. In 1854 he became one of the secretaries of the Royal Society. This post he accepted in part because, although he held the most honorable professorship of mathematics in the world, the University of Cambridge was not "able to assign some endowment to the Lucasian Chair" until 1859, and the Royal Society paid each of its two secretaries £100 per annum. During Stokes' long tenure as secretary on the mathematical and physical side, he spent an immense amount of time in improving the papers of other investigators and, although his own research suffered, he greatly raised the standard of the work of others. He gave continual advice on scientific matters, mainly meteorological, to various public bodies. A large amount of work on the optical properties of glasses of known composition, prepared by W. Vernon

Harcourt, led to few and brief publications but had an immense influence on the construction of optical instruments, notably the great telescopes being built at this time by Sir Howard Grubb.

Many of Stokes' contributions in later years were appended to the papers of other authors, the last two being on the mathematical aspects of the most efficient processes of fractional crystallization, and on the best method of interpolation in wave-length measurements, both attached to papers by Sir William Crookes, one on rare earths and the other on the ultraviolet spectrum of radium. When the latter paper was published Stokes had just died, and the Victorian age, both politically and scientifically, was over.

There is no doubt that, in spite of these numerous scientific preoccupations, Stokes in later life became more and more of a theologian, as did Newton. Here he was much less successful than as a physical scientist. The peculiar compartmentalization of his mind, partly a matter of temperament, partly of early upbringing, which his friends noticed, led him to avoid most of the religious doubts besetting his friends and colleagues. But for the same reason, when he attempted to reconcile science and religion in order to answer these doubts, the results were not satisfactory. He felt that the greatest stumbling block in the minds of his agnostic friends (T. H. Huxley was among them) lay in the doctrine of hell, which could hardly be reconciled with belief in a God of infinite goodness. He, therefore, read the New Testament carefully and found no unequivocal evidence for the belief in eternal damnation. The doctrine of the unconditional immortality of the soul he regarded as a piece of unwarranted Platonic metaphysics; conditional immortality would be bestowed on the righteous, the souls of the wicked would merely cease to exist. He felt that this reinterpretation should do much to heal the rift between nineteenth-century science and orthodox Christianity. Actually, it only led people to say that he held, as Sir Ellis Minns puts it, "peculiar views about the life after death and wrote tracts about the subject."

Stokes gave two series of Gifford Lectures, published under the title *Natural Theology*. These were largely restatements of the general point of view of the Bridgewater Treatises. The rejection of generality in favor of the physically real, which had been such a source of strength in his early hydrodynamic researches, now became a source of weakness of which he was conscious. "I think a too minute dissection of the processes of nature, with a view to bring out evidences of design, should rather be held in check by a due feeling of reverence, and that the free indulgence of it may tend to make our ideas of the Author of nature too anthropomorphic."[5] Yet Stokes did just this when he really approached the heart of the matter. He knew that the colors of the rainbow have a full physical theory behind them, and had once pointed out to Millais, the Royal Academician, that he had painted a rainbow inside out. Stokes knew that he found the rainbow beautiful.[6] He also found plants and animals beautiful and he knew that much of the beauty could be due to natural selection producing patterns of utility. He hoped that some of the beauty was put there to please the eye. He knew that all patterns may have a utilitarian explanation, but he could not see in his passion for an explanation that the beauty is independent of the explanation, and that the explanation itself has beauty independent of the explanation of the explanation . . . Just for a few lines, one feels such a revelation coming, but the sentences veer off and we are back in the world of the Earl of Bridgewater's legacy.

Perhaps it does not matter that Stokes' theology seemed peculiar to his friends and that it is dead today. His daughter wrote, "We thought that he had received some special illumination towards the close of his life." The words in which he might have attempted to express that illumination doubtless failed him, as words have failed many others. For us, it must be enough to know that he was a great and good man who shed light on many things which previously had been in the dark.

5. *Natural Theology: The Gifford Lectures Delivered before the University of Edinburgh in 1893* (London, A. and C. Black, 1893), p. 94.

6. Ibid, p. 99f.

CAMBRIDGE REMEMBERED

For those unhappy enough not to know the place, it must first be observed that a river flows through the town. Originally there were two settlements, one on the northwest side of the river, one on the southeast. Both were occupied in Saxon times.

To those of us who lived south and east of the river, the transpontine northern settlement had almost the feeling of an alien country. We thought that the sun shone there less often and with feebler rays. The inhabitants were perhaps giants, superior beings in an obviously unpleasant and inferior land.

Originally the southern settlement had lain a little east of the marshes that bounded the river. It was in fact on an artificial island, the river on one side, a ditch on the other. Gates opened on the south and east, and on the north was a bridge to the northern settlement where the Normans built a small castle. In medieval times various conventual and collegiate establishments arose, mostly outside the ditch or on land reclaimed between the town and the river.

As in more modern times the town grew, finally achieving the dignity of a city, farm lands on the south, west, and north became respectable residential areas, the *superego* of the settlement. To the east there was an area of industrial development of small nineteenth-century slum streets whose inhabitants worked on the railway, in clay pits or brickyards, or in the few small industries. These people, when one first in early childhood ran across them, were people. Class distinction meant nothing to a child. Later some of their traits were seen to elicit adult disapproval. They frequented fish-and-chips shops; they got drunk; they had daughters called Hannah and Anna, spelled Anna and Hannah. They often were patently unwashed. This region, which had sprawled over the former estate in the Augustinian Priory of Barnwell, was the *id*, the forgotten unconscious, of the place. The *ego* radiated from the flat area in the middle called Market Hill, where butter was sold by the yard.

The small boys in the areas defined as the superego went to one of two schools. One was originally founded as a choir school for King's College Chapel, but it took paying pupils as well as the choristers who were chosen for their voices from all over England, had a free education, but had to remain behind to sing at Christmas. The other school was named, for no ascertainable reason, for Saint Faith. Whether the patroness was Saint Faith of Agen, or the quite mythical sister of Saint Hope and Saint Charity, no one ever inquired.

The area to the south of the town was underlain by Pleistocene river gravel. Whenever a gravel pit was dug there was a chance of discovery of hippopotamus or elephant bones. The area north and west of the river lay on clay. It was supposed to be less healthy to live on than the gravel. However, most of the available land to the south could be obtained only on a 99-year lease, and one might conceivably live that long and see Trinity College dispossessing one. The transpontine land to the north and west was largely freehold. The boys from the clay freehold homes went to the King's Choir School. They developed the kind of superiority that one otherwise associated with Oxford, and always seemed to carry umbrellas. The boys from the gravelly leasehold homes went to Saint Faith's and tried hard to beat the King's Choir School at football and cricket. How the girls got educated was something of a mystery, but there are distinguished women now living who prove conclusively that the process did occur.

One home in the northern land typified that area. In it grew up the present Archbishop of Canterbury, and his older brother,

Frank Ramsey, who, dying at 26, left an indelible mark on logic and the foundation theory of mathematics and science. Once at a children's party, a guessing game called "clumps" was in progress. Frank was sent into the hall to arrange with a like delegate from the opposing side the object to be guessed. His choice, which prolonged the game greatly, was the left horn of a dilemma.

This dichotomy of southeast and northwest probably did not affect the adults. The men of the superego areas were almost all associated with the University. They were for the most part practically invisible. One might, when out walking with a grandmother, meet Mrs. Skeat, Mrs. M'Taggart, Mrs. Hort, or Mrs. Frazer, and later discover these names were famous ones that the ladies in question had acquired. There were also the Misses Miller who went out driving in a victoria; their father had invented the Miller indices that make crystallography, and so much of modern science, possible. There were a few exceptional visible men that every well-brought-up child knew by sight, at a respectful distance. The Reverend Dr. Bonney, author of *Volcanoes: Their Structure and Significance*, bent in a right angle, still walked, in his old-fashioned long clerical coat, up the Trumpington Road. Everyone was familiar with the sight of Sir J. J. Thompson, a slight figure who seemed to be covered with electrons, each individual hair sticking out from his moustache or from under his hat repelling all the others. Much later Professor G. E. Moore, destined to become another Cambridge O M, was equally well known, pushing his twins in a perambulator. Throughout the whole period the place was inhabited by genius, possibly more significant intellect per thousand inhabitants than any place has had before or since.

There was at this time in Cambridge an undergraduate called Lillie. He has left a little-known but wonderful record of some of the remarkable men of the time in a series of caricatures. Nothing more need be said when his drawings are available (Frontispiece).

It was perhaps easier to be aware of houses than of their inhabitants. There were for instance several houses inhabited by the sons of

Charles Darwin, all knighted. What they looked like nobody under late adolescence probably knew. Once, however, it was reliably reported that Sir George and Lady Darwin had come to dinner. Two small boys waited at the top of the stairs watching the guests go in to dine. Lady Darwin, presumably the female guest of honor, was in a magnificent cerise dress. In a few minutes the party was locked in the dining room and the lights turned out at the main switch. What then happened cannot be part of this essay, for it is not among the available memories, or fantasies, of the place.

The women of the area, mostly the wives of less distinguished men than those that have been named, were no doubt more diversified than appeared. They were for the most part not learned. Miss Edith Saunders was a conspicuous exception. She was said to have bought a garden-party dress in the 1890s and to have worn it at the Newnham College garden party each year until her death nearly half a century later. At about the same time she bought the dress, she had before her a group of seedling stocks, part hairy and part smooth and in the proportions that would be expected today from Mendel's laws of inheritance. Unhappily she did not count the two kinds of seedlings, and Mendel had to be rediscovered by three workers on the continent. Far more remote, but far less terrifying, were the Heavenly Twins, Mrs. Gibson and Mrs. Lewis, whose legal possession of a fortune was said to have depended on their never sleeping under different roofs while they both lived. Both widowed in middle life, they traveled extensively and, visiting Mount Sinai, became among the greatest of Syriac scholars. Small old-fashioned ladies, by their romantic history and the nature of their studies, they linked the present with every imaginable human past. It is fascinating to think of them, photographing their great find of a very ancient palimpsest gospel, largely erased and covered by a later unimportant Syriac text, with the cumbersome equipment that they had in the Middle East in 1892.

In general, however, one only really recognized one's mother's friends. They seemed essentially nonvisual, dressing in brown, gray and dark green, uncertain procryptic colors that faded into the landscape. They themselves were large boned and vigorous, with a strictly constitutional passion for female suffrage. When they rode through the town on bicycles it was difficult not to absorb the impression that men were responsible for all the evil in the world. Happily, at that time in Cambridge they seldom came up against the more feminine kinds of intellectual women, whose existence they helped to make possible. Between such women and men, the feminist pioneers would have been crushed. Actually they felt very secure, traveling to London for meetings, knowing few people outside the University, and regarding, when occasionally they thought of them, the ladies of the scattered large country houses in the county as incurably frivolous. In fact it appeared quite natural that Alyne, wife of Sir John de Creek of Westley Waterless, a place so inaccessible that no one seemed ever to have been to it, should have left in her monumental brass the finest record of *haute couture* circa 1325 known in England. Ladies of this sort obviously were not very satisfactory; some of them indeed were definitely "fast," a word that was difficult to pronouce without hesitation.

The undergraduate members of the university made little impression except at boat-race times; their more exciting qualities were reflected in the sight of proctor and bulldogs, and, if one was incredibly lucky, in a mock funeral. The proctors, who maintained university discipline in the streets, walked around in cap and gown attended by two stalwarts in uniforms and top hats, carrying keys and a volume of the University statutes. The chief duties of the proctors were to apprehend and fine any student found not wearing academic dress after dark. Occasionally they could make a much more satisfactory score by fining a really senior member of the University for a comparable if more ceremonial offense, as when the Master of Christ's was found outside his college on Easter Day in a black gown, when, if he were to wear academic dress at all that day, it should, since he held an Sc.D., have been the full dress scarlet.

For a child, the best method of encountering a proctor with

bulldogs was to be taken for a walk in King's Parade on Sunday afternoon, when such dignitaries were leaving the University Sermon. Since, however, this meant a Sunday afternoon walk in the town, which was considered unhealthy, it could rarely be arranged.

A mock funeral took place when an undergraduate was sent down from the university for some really serious offense. A funeral procession of various kinds of cabs followed the "corpse" to the station. The practice was forbidden well before the local extinction of the carriage horse, on which it depended. Mock funerals were not the sort of entertainment parents chose for their children; they raised too many awkward questions, but everyone must at least have heard of the more spectacular ones, such as that of T. E. Hulme. Hulme, killed in the ensuing war, returned to Cambridge in book form in the twenties. Transported to America, he has been with us, for good or ill, ever since.

As well as dons and undergraduates, University society included a few ghosts, but they were very hard to meet. Two colleges, Corpus Christi and Jesus, seem to have been particularly haunted. There was also a good ghost in a house near the site of Barnwell Priory, and a small, furry, quite unacademic ghost, through which one could stick penknives as it ran up the curtains of an ordinary private house in Trumpington. The really serious ten-year-old parapsychologist, like his older colleagues today, had a very adverse and negative effect on such apparitions, which seemed to favor the unintellectual hearty type who had no idea of evidence and were possibly not above practical joking. In spite of the difficulty of the subject, the study of ghosts had a perennial fascination.

The natural history was easier to study than the supernatural. Every autumn one could find the large fawn-and-chestnut-colored caterpillars of the goat moth crossing the road, looking for sites into which to burrow and become chrysalises. They were a perennial challenge and irritation; obviously common, they burrowed in wood for most of their lives and became visible only for a few days. The adult moth never appeared, and the problem of rearing the larva through metamorphosis, though attempted over and over again, seemed too difficult for a small boy. Equally exciting were the bizarre caterpillar of the puss moth, the various species of hawk moth—poplar, lime, and eyed—and the great green metallic musk beetle that sometimes was seen sitting on arrowhead leaves growing from backwaters of the river. Coe Fen and Sheeps Green, adjacent rough pastures with ditches and ponds, now alas sadly disfigured, yielded newts and sticklebacks; occasionally one saw the kingfisher, the most spectacular of all ordinary British birds, darting along a stream, a double line of sapphire and chestnut on the southern side of Coe Fen. Further into town, where the river runs through the loveliest artificial landscape man has created, Daubenton's bat used to sleep under the spans of the bridges and hawk gnats up and down over the water at dusk. For some unknown reason, time has laid a heavy hand on this attractive animal; bats hawk up and down the Arno, but no longer much on the Cam.

Much more obscure, and so in some ways more fascinating, were the animals that one knew some learned men studied, for the town was full of amateur naturalists as well as having its academic professional zoologists. Francis Jenkinson, the tall wiry University Librarian, collected flies; he used to tell how he heard a rare and large fly above his head, seized a tennis racket from a startled young lady walking by, and brought down his prize with a leap and a high backhanded stroke. F. H. H.Guillemard had collected birds of paradise as a young man in New Guinea, carefully covering the exposed flesh as he skinned them with small pieces of toilet paper, and later he had made a point of attending every sale of a great auk egg or skin held in Britain.

More professional, though a man of independent means, George P. Bidder, grandson of a well-known calculating boy and himself mathematically inclined, appeared as an invalid with magnificent beard and whiskers, seated under a tree in his extensive garden. As a young man he had bought Parker's Hotel in Naples, to avoid being got out of bed, so the story ran, when the property was being prepared for sale by auction. Though he became progressively more nocturnal and published relatively little, the half-dozen papers that he did complete, dealing mainly with the biology of sponges, have a fascination to the zoologist, and one is regarded as a classic. Late in life he taught an advanced course on the sponges; he never managed to finish a consideration even of the calcareous species, though these are greatly outnumbered by forms with siliceous or organic skeletons. The course was, however, a series of footnotes on scientific civilization. To be a good zoologist, Bidder said, one must know everything except possibly irregular Greek verbs. Once in the 1920s at a scientific meeting, the subject of the biology of size was introduced by J. B. S. Haldane with the comment that of the three founders of the study, Galileo, D'Arcy Thompson, and Bidder, two were fortunately present in the room. These two survivors were also greatly gifted as writers of English prose: "The cavity of the Hexactinellida is no pressure-chamber: it is even perforated to let the onflowing water sweep out that which is befouled. Food is brought to them, waste is taken away. For them in their eternal abyss, with its time-like stream, there is no hurry, there is no return. Such an organism becomes a mere living screen between the used half of the universe and the unused half—a moment of active metabolism between the unknown future and the exhausted past."

It is such a moment in the ocean of the human intellect that this essay is written to record, seen by a child who could know little of what was implied, but, being human and not a glass sponge, was able in the memory to stay the current of time.

INDEX

Printed by Libri Plureos GmbH in Hamburg,
Germany

9 780313 200984